Guidance on
addressing child labour
in fisheries and aquaculture

Published by
the Food and Agriculture Organization of the United Nations
and
the International Labour Organization

International
Labour
Organization

Foreword

About 60 per cent of all children engaged in child labour are active in the agriculture sector, including fisheries and aquaculture, forestry and livestock. In fisheries and aquaculture, children engage in all types of activities, from catching fish to repairing nets or processing fish, often in a way that is incompatible with school attendance and hazardous to their health.

Child labour is defined as work of children who are too young for the type of work they do, work that interferes with their schooling and, as applies to all children under 18 years of age, work that risks harming their health, safety or morals. Of course, not all activities children engage in are child labour. Some activities may stimulate their development as they allow them to acquire precious skills and contribute to their survival and food security. These activities can be beneficial as long as they are not hazardous, not undertaken for long hours and do not interfere with school and learning.

Work hazardous to children and preventing their physical, mental, spiritual, moral or social development is not acceptable under any circumstance. That is why it is so important to understand what we mean by child labour, why it happens, and what can be done to eliminate it.

Child labour hampers societal progress. It affects children's development and hence their capability to work and productivity when becoming adults. It also poses a threat to business sustainability. Respect for human rights, including the rights of the child and fundamental rights at work, is an important element in the choice of business partners. In a globalized world, the presence of child labour may thus exclude producers from lucrative value chains and new market opportunities. This also applies to fisheries and aquaculture.

For all these reasons it is especially important to reach out to family and small-scale fishing and aquaculture operators to help raise their awareness of what constitutes child labour and hazardous work. This will be a first step towards improving the occupational safety and health of all those working in the sector, and to ensure that children are not involved in activities that are harmful for them, while enhancing availability and access of relevant school and vocational training.

The international legal framework to address child labour, based on the Minimum Age Convention 1973 (No. 138) and the Worst Forms of Child Labour Convention 1999 (No. 182), is still not adequately applied and enforced in many contexts and child labour remains prevalent especially among informal, small-scale informal fisheries and aquaculture enterprises.

FAO and the ILO have therefore joined forces. Together with other partners in the framework of the International Partnership for Cooperation on Child Labour in Agriculture launched in 2007, they work to better address child labour in agriculture. FAO and the ILO share a commitment to support people-centred, sustainable development and fair and inclusive globalization. The joint strategy seeks to complement and strengthen the work currently undertaken by both organizations, on labour rights, agriculture and rural development, and the promotion of decent work in rural economies.

The present document is a response to one of the recommendations of a Workshop on Child Labour in Fisheries and Aquaculture held in Rome in 2010, which requested FAO and ILO to develop guiding principles for policy makers, organizations of fishers, fish farmers, fish workers and employers, and other sectoral institutions, development practitioners, and the general public to address child labour in the fisheries and aquaculture sector.

We hope that this document will provide insights and entry points for different stakeholders to find ways to break out of the vicious cycle of child labour and poverty.

Jomo Kwame Sundaram
Assistant Director-General, Economic and Social Development Department
Food and Agriculture Organization of the United Nations (FAO)

Árni M. Mathiesen
Assistant Director-General, Fisheries and Aquaculture Department
Food and Agriculture Organization of the United Nations (FAO)

Moussa Oumarou
Director, Governance and Tripartism Department
International Labour Organization (ILO)

Preparation of this document

This document – *Guidance on addressing child labour in fisheries and aquaculture* – has been prepared within the framework of the collaboration between the Food and Agriculture Organization of the United Nations (FAO) and the International Labour Organization (ILO) on decent work and child labour in the food and agriculture sector.[1] It responds to a need to better understand and address child labour in fisheries and aquaculture drawing on available information and material.

The text is based on the outcomes and recommendations of a workshop held in 2010, the FAO Workshop on Child Labour in Fisheries and Aquaculture in Cooperation with ILO (FAO, 2010a). Inputs were also provided in the FAO–ILO Workshop on Capacity Development on Child Labour in Agriculture (including Fisheries and Aquaculture), organized in Salima, Malawi, in May 2011 (IPCCLA, 2011). A preliminary version of the guidance document was prepared in 2011 and this document incorporates comments on and inputs to the earlier draft.

The document was prepared by Lena Westlund, FAO consultant, with substantial support from FAO and ILO colleagues (in particular Jacqueline Demeranville, Nicole Franz, Carlos Fuentevilla, Anne-Brit Nippierd, Yoshie Noguchi, Deepa Rishikesh, Bernd Seiffert, Paola Termine, Faustina Van Aperen, Brandt Wagner and Rolf Willmann). Comments made by Katrien Holvoet (FAO Fisheries and HIV/AIDS in Africa programme) and Tomoko Horii and Bruce Grant (UNICEF, Malawi) are also gratefully acknowledged. We would like to thank Ruth Duffy for the final editing and the team at International Training Centre of the ILO (ITCILO) for layout. The document's preparation was made possible thanks to the United States Department of Labor (USDOL)'s support to the International Programme on the Elimination of Child Labour (IPEC) of the ILO and to the International Partnership for Cooperation on Child Labour in Agriculture, and thanks to Swedish funding through the FAO Multipartner Programme Support Mechanism (FMM).

For additional information on child labour in agriculture you are invited to visit the following websites:

FAO–ILO webpage on child labour in agriculture: www.fao-ilo.org/fao-ilo-child/en

International Programme on the Elimination of Child Labour (IPEC) of the ILO: www.ilo.org/ipec

International Partnership for Cooperation on Child Labour in Agriculture: www.fao-ilo.org/fao-ilo-child/international-partnership/en/

Contact: ipec@ilo.org or VG-SSF@fao.org

---

1   See www.fao-ilo.org/.

Abstract

Child labour is a major concern in many parts of the world and it is estimated that there are some 215 million child labourers worldwide. Aggregate data indicate that about 60 per cent of child labourers – that is over 129 million children – work in agriculture, including fisheries and aquaculture. While there are limited disaggregated data on child labour specifically related to fisheries and aquaculture, case-specific evidence points to significant numbers. Children engage in a wide variety of activities in capture fishing, aquaculture and all associated operations (processing, marketing and other post-harvest activities), as well as in upstream industries including net making and boatbuilding. Children also perform household chores in their fishing and fish farming families and communities. When child labour is used as cheap labour to cut fishing costs, not only is harmful to the children, it may also have a negative effect on the sustainability of the fishery activity. Child labour appears to be particularly widespread in the small- and medium-scale sectors of the informal economy where decent work is poorly organised or absent.

Although there is a widely ratified international legal framework to address child labour, – comprising ILO Conventions and other agreements, laws are effective only if they are applied and enforced, with incentives to ensure compliance. Addressing child labour is rarely high on the national agenda of social dialogue, legislation review and institution building. Its elimination is difficult because it is part of production systems, is nested in the context of poverty and relates closely to social injustices. Communities and institutions are often not fully aware of the negative individual and collective social and economic consequences of child labour. Practical and realistic pathways for improving the current situation and community engagement and buy-in are essential for successful results.

More information on child labour is needed to raise awareness at all levels. A critical first step towards eliminating child labour, in particular its worst forms, is to understand what constitutes hazardous work and what tasks and occupations are acceptable for children above the minimum legal age for employment. Not all activities performed by children are child labour. Convention on Minimum Age, 1973 (No 138), and Convention on the Worst Forms if Child Labour, 1999 (No. 182), define child labour on the basis of a child's age, the hours and conditions of work, activities performed and hazards involved. Child labour is work that interferes with compulsory schooling and damages health and personal development.

Concerted efforts are needed to effectively address child labour with multistakeholder participation and involving governments, development partners, non-governmental organizations (NGOs), employers' and workers' associations and other socioprofessional organizations, the private sector and communities (including children and youth). By applying holistic, participatory, integrated and feasible approaches, a better life for millions of children can be created.

Contents

List of tables

List of figures

List of boxes

Abbreviations and acronyms

CAMCODE	Cambodia Code of Conduct for Responsible Fisheries
CCT	Conditional Cash Transfer
CCRF	Code of Conduct for Responsible Fisheries
CGIAR	Consultative Group on International Agricultural Research
CLM	Child Labour Monitoring
CONITPA	National Coordination of Port and Waterway Labour Inspection (Brazil)
CRC	Convention on the Rights of the Child
EJF	Environmental Justice Foundation
FAO	Food and Agriculture Organization of the United Nations
FMM	FAO Multipartner Programme Support Mechanism
GAWU	General Agricultural Workers Union (Ghana)
GESAMP	Joint Group of Experts on the Scientific Aspects of Marine Environmental Protection
ICLS	International Conference of Labour Statisticians
ICSF	International Collective in Support of Fishworkers
IFAD	International Fund for Agricultural Development
IFAP	Industrial Foundation for Accident Prevention
IFPRI	International Food Policy Research Institute
ILO	International Labour Organization
IMO	International Maritime Organization
IOM	International Organization for Migration
IPCCLA	International Partnership for Cooperation on Child Labour in Agriculture
IPEC	International Programme on the Elimination of Child Labour (ILO)
ITCILO	International Training Centre of the ILO
IUF	International Union of Food, Agricultural, Hotel, Restaurant, Catering, Tobacco and Allied Workers' Associations
JFFLS	Junior Farmer Field and Life Schools
LSMS	Living Standard Measurement Surveys
MDG	Millennium Development Goal
MLVT	Ministry of Labour and Vocational Training (Cambodia)

NAP	National Action Plan
NGOs	Non-Governmental Organizations
NPA	National Plan of Action (Cambodia)
OHCHR	Office of the High Commissioner of Human Rights
OSH	Occupational safety and health
PPE	Personal Protective Equipment
REC	Regional Economic Communities
RFB	Regional Fishery Body
SAR	Search and Rescue
SEAGA	Socio-Economic and Gender Analysis
SECTOR	Sectoral Activities Department (ILO)
SEWA	Self-Employed Women's Association (India)
SIMPOC	Statistical Information and Monitoring Programme on Child Labour (ILO)
SPF	Strategic Planning Framework (Cambodia)
SSF	Small-Scale Fisheries
STCW-F 1995	International Convention on Standards of Training, Certification and Watchkeeping for Fishing Vessel Personnel, 1995
TFD	Theatre for Development
UNDP	United Nations Development Programme
UNESCO	United National Educational, Scientific and Cultural Organization
UNICEF	The United Nations Children's Fund
UNODC	United Nations Office on Drugs and Crime
USA	United States of America
USDOL	United States' Department of Labor

Executive summary

Child labour continues to be a major concern in large parts of the world. It has been estimated that there are some 215 million child labourers worldwide and that approximately 60 per cent (129 million) of them work in agriculture, including fisheries and aquaculture. About 59 per cent (or 70 million) of all children aged 5–17 in hazardous work are in agriculture. Addressing child labour is difficult because it is intertwined with poverty, social injustices and the very structure of production systems. Reliable data are often lacking, particularly in the fisheries and aquaculture sector.

This *Guidance on addressing child labour in fisheries and aquaculture* provides information and analyses current issues in order to improve the understanding of the nature and scope, causes and contributing factors, and consequences of child labour in fisheries and aquaculture. It aims to contribute to the prevention and elimination of child labour in the fisheries and aquaculture sector by assisting governments and development partners to better define and classify child labour in fisheries and aquaculture, to mainstream child labour considerations in relevant development and management policies, strategies and plans, and to take action against child labour. The document is directed at government officials and their development partners, organizations of fishers, fish farmers, fish workers and employers, and other sectoral institutions, the private sector and other stakeholders, in the formal and informal economy. It gives guidance on how to find practical pathways to address child labour and provide support to fishers, fish farmers and fish workers and their communities, in particular in the small-scale sector.

Economic activities and household chores performed by children are not necessarily child labour. While child labour by definition is unacceptable and should be abolished – in particular the worst forms of child labour – there are activities that are not harmful to children and may even be beneficial. Child labour is defined by the ILO on the basis of a child's age, the hours and conditions of work, activities performed and hazards involved. Child labour is work that interferes with compulsory education and damages health and personal development.

In the fisheries and aquaculture sector, children engage in a wide variety of activities, both in capture fishing and aquaculture and in all associated down and upstream operations, for example, processing, marketing, net making and boatbuilding. Children also perform household chores in their fishing and fish-farming families and communities. Child labour appears to be particularly widespread in the small- and medium-scale sectors of the informal economy. When child labour is used as cheap labour to cut fishing costs, not only may it be harmful to children's development, it can also have a negative effect on the sustainability of the fishery activity, and the economy.

While there is an international legal framework for addressing child labour consisting of ILO Conventions and other agreements, the reality is that many such tools are yet to be translated into national legislation and implemented. Fisheries and aquaculture is a particularly neglected sector that offers poor protection to men and women workers. A fundamental international commitment with regard to children is the UN Convention on the Rights of the Child, 1989. The Minimum Age Convention, 1973 (No. 138) and the Worst Forms of Child Labour Convention, 1999 (No. 182) are the key international instruments for child labour. The FAO Code of Conduct for Responsible Fisheries and the Work in Fishing Convention, 2007 (No. 188) are relevant instruments and guidelines specific to fisheries and aquaculture.

Governments need to adopt national policies and put in place legal and institutional frameworks to address child labour. However, laws are only effective if implemented and enforced, and incentives are required to ensure compliance. Community awareness and buy-in are essential for successful results. Stakeholder engagement and participation are particularly important in the informal economy. Actors in the small-scale sector require support to find practical and relevant solutions. When migration is a common livelihood strategy, as is the case in many small-scale fishing communities, regional and cross-border collaboration may be required.

A system of classification is needed to distinguish between acceptable activities that are adequate for children, and child labour or even hazardous child labour (worst forms of child labour). It is essential to first understand what constitutes hazardous work. Criteria for defining hazardous work onboard a fishing vessel could include hours at sea, weather conditions, type of gear used, related work processes, need for diving and general working (and living) conditions onboard the vessel. In the post-harvest sector, in boatbuilding and in aquaculture, there are other potential hazards, including exposure to smoke (when smoking fish), the noise level (in boatbuilding) or use of toxic substances (in aquaculture).

There is a lack of information on child labour regarding the different causes and consequences for boys and girls. It is therefore necessary to integrate data collection needs into existing information systems and processes, and carry out specific assessments in collaboration with stakeholders. It would then be possible to raise awareness at all levels, as well as develop cross-sectoral capacity in support of policy coherence; fisheries and aquaculture policies and programmes should incorporate child labour concerns, and child labour strategies should take into consideration the characteristics of fisheries and aquaculture.

Action against child labour comprises (i) prevention, (ii) withdrawal and (iii) protection:

- **Prevention** is the most important approach for addressing the root causes of child labour and achieving long-term sustainable results. It includes poverty-focused, participatory and integrated programmes aimed at turning the vicious cycle of poverty and child labour into a virtuous cycle leading to sustainable development (see Figure 1). Childcare and education must be made adequate, relevant, affordable and available, and incentives are required to ensure that children attend school, for example, school feeding programmes or (in some cultural contexts) separate schools for girls. Technologies and practices to reduce the demand for child labour can ensure the sustainability of preventive strategies. It is important to raise awareness of children's rights – also among children themselves – change attitudes through education and promote corporate social responsibility.

- **Withdrawal** is an urgent intervention and can be necessary to rescue and rehabilitate children engaged in child labour and worst forms of child labour. Close community participation and collaboration are important for sustainable results.

- **Protection** for children above the minimum legal age for employment (normally in the 15–17 years age group) leads to improved working conditions, and to transforming hazardous work into youth employment opportunities. Onboard fishing vessels, protection can be improved by making available and using life jackets.

To implement actions, it is necessary to have entry points, as well as partners and tools that are suitable and which work in the particular local context:

- **Entry points** include overall occupational safety and health (OSH) assessments and improvement actions, and – in fishing – safety at sea.

- **Partners** should be at national and local level, including line ministries and government agencies (needed for an integrated institutional approach), socioprofessional organizations and employers' and workers' organizations.

- **Tools** already exist. They need to be more widely applied to eliminate child labour. They include risk assessments and checklists for evaluating child labour in a particular situation, policy, legal and institutional analyses, national action plans drawn up in participatory workshops, reports and events to increase knowledge of child labour, and – at community level – working methods and communication tools, such as participatory assessments, radio and TV programmes, public and village meetings, and Theatre for Development (TFD).

Concerted efforts are needed to effectively address child labour. This requires the involvement of governments, organizations of fishers, fish farmers, fish workers and employers, and other sectoral institutions, their communities, development partners and non-governmental organizations (NGOs). Each group needs to take on important roles, responsibilities and tasks. Through participation of different stakeholders and application of holistic, integrated and feasible approaches, it is possible to create a better life for millions of children.

Laos, girl with a net in the water
©FAO

Introduction

Background

Child labour continues to be a major concern in many parts of the world. In 2008, some 60 per cent of the 215 million boys and girls estimated to be child labourers around the globe were engaged in agriculture, including fisheries, aquaculture,[2] livestock and forestry (ILO, 2010a). This work not only interferes with schooling and is harmful to personal development, but many children are involved in occupations or activities that may threaten their health and lives. Children perform work prohibited under international conventions and/or national legislation, creating a situation that is a menace to themselves and to sustainable development.

Tackling child labour is a complex task; child labour is entwined in poverty and social injustices and cannot be addressed in isolation. Moreover, some types of work are not harmful and can even be beneficial for children. Indeed, while it is relatively easy to identify and agree to eliminate the "worst forms of child labour", including hazardous work,[3] the distinction between activities that are acceptable and age-appropriate and child labour (by definition engagement in activities that are detrimental for a child's development and well-being) is not always clear and assessments can be muddled by local and traditional practices and beliefs. This is often the case in family-based undertakings and in the informal economy, where child labour is particularly common (for example, in small-scale agriculture, fisheries and aquaculture). It is necessary to work closely with communities to carefully analyse existing situations and raise the awareness and understanding of child labour issues and find practical solutions. By taking a participatory approach while promoting the application of existing Conventions, legislation and guidelines, child labour can be addressed directly and integrated into broader policies and programmes. Improvements are possible; in fact, available data show that the total number of child labourers in the world has declined since the year 2000.[4]

Information on child labour in fisheries and aquaculture is limited, and data on child labour in agriculture are generally not disaggregated by subsector. Nevertheless, case studies and specific surveys indicate that the numbers are significant. Children are engaged in a large variety of activities. They perform non-economic tasks (household chores) and also carry out economic activities as part of family enterprises,

---

2   Aquaculture is the farming of aquatic plants and animals in marine, freshwater or brackish water environments.
3   Hazardous child labour is work that is likely to harm a child's health, safety or morals and is a worst form of child labour (ILO IPEC website). See also Box 1 in chapter 1, What is child labour?.
4   See section 1.4, Data on child labour.

as unpaid family workers, self-employed or employed by others. They work onboard fishing vessels, unloading catches, preparing nets and baits, feeding and harvesting fish in aquaculture ponds, and sorting, processing and selling fish.

There are a number of factors that determine whether a task is considered acceptable and adequate, child labour or a worst form of child labour, such as hazardous work. With the support of initiatives, including the International Partnership for Cooperation on Child Labour in Agriculture, launched in 2007,[5] the knowledge base and guidance on how to classify and address child labour in agriculture have improved. As part of enforced efforts to eliminate child labour in agriculture, especially hazardous child labour, the ILO and FAO are collaborating to develop guidance on policy and practice. In April 2010, FAO held a Workshop on Child Labour in Fisheries and Aquaculture, in cooperation with ILO (FAO, 2010a). The development of this document is a direct follow-up to the recommendations of the workshop and is part of a broader long-term commitment by FAO and the ILO to address child labour and promote decent work in the wider agriculture sector, including fisheries and aquaculture. There is also extensive collaboration among partners as part of ILO's International Programme on the Elimination of Child Labour (IPEC). IPEC was created in 1992 with the objective of progressively eliminating child labour, by strengthening the capacity of countries to deal with the problem and promoting a worldwide movement to combat child labour. In 2010–11, IPEC provided technical assistance to 102 member States and maintained operational activities in 88 countries.[6]

Purpose and structure of this document

The objective of this document is to provide guidance for the elimination of child labour in the fisheries and aquaculture sector. It aims to assist governments and development partners to better define and classify child labour in fisheries and aquaculture, to mainstream child labour considerations in relevant development and management policies, strategies and plans, and to effectively address child labour in fisheries and aquaculture. The document sets out to improve understanding of the nature and scope, causes and contributing factors, and consequences of child labour in fisheries and aquaculture. It includes examples of good practices relative to priority actions to be taken by governments, their development partners and sector representatives. It recognizes the particular and often difficult situation, including poverty and marginalization, of small-scale fishers, fish farmers and fish workers and their communities and adopts a practical approach.

The target audience of the guidance document includes government officials and their development partners; those involved in the fisheries and aquaculture sector; and those operating in areas where child labour occurs and where a better understanding of issues particular to fisheries and aquaculture is needed. This guidance is also directed at organizations of fishers, fish farmers, fish workers and employers, and other sectoral institutions in the formal and informal economy. Concerted efforts are required to address child labour; all concerned parties need to be involved and take action.

---

5 Current members of the International Partnership for Cooperation on Child Labour in Agriculture (IPCCLA) are the International Labour Organization (ILO), Food and Agriculture Organization of the United Nations (FAO), International Fund for Agricultural Development (IFAD), International Food Policy Research Institute (IFPRI) of the Consultative Group on International Agricultural Research (CGIAR), and the International Union of Food, Agricultural, Hotel, Restaurant, Catering, Tobacco and Allied Workers' Associations (IUF).

6 See www.ilo.org/ipec/programme/lang--en/index.htm.

The scope of the document is global, applying to all the different fisheries and aquaculture subsectors – small- and large-scale capture fisheries, aquaculture and post-harvest activities. However, because of the higher prevalence of child labour in the informal economy and the particular circumstances and poverty context that are often part of the reality of small-scale capture fisheries in developing countries, emphasis has been placed on this part of the sector.

Following this introductory chapter, **Part 1** gives an overview of the concepts, definitions and recent data relevant to child labour. It presents the characteristics of the fisheries and aquaculture sector and the different subsectors, and discusses child labour within this context. It provides examples of what children do in fisheries and aquaculture and the related risks and hazards.

Part 2 begins with an overview of existing legal and policy frameworks relevant to child labour, while emphasizing the importance of community engagement and awareness raising. It presents good practices and gives suggestions on how to classify, mainstream and address child labour in fisheries and aquaculture.

This guidance document can be used to look up specific information on different topics; not all readers may need to read the full text. **Table 1** provides suggestions on how to find answers to different questions. All readers, however, are strongly recommended to refer to the end of the document and the section entitled **Summary of recommendations**, including recommendations for different stakeholder groups.

TABLE 1: STRUCTURE OF THIS DOCUMENT

Question or point of interest	Where to look in this document
I know the fisheries and aquaculture sector but need to know more about child labour.	**Part 1/chapter 1** *(What is child labour?)* defines and gives general information on child labour. In **Part 1/chapter 3** *(Children in fishing and aquaculture)*, acceptable activities and child labour in fisheries and aquaculture are discussed. Additional information on developmental differences between children and adults can be found in **Appendix 1**.
I am not familiar with the fisheries and aquaculture sector and would like to get an overview of key issues.	**Part 1/chapter 2** *(The fisheries and aquaculture sector)* describes sector-specific characteristics, challenges and opportunities, as well as general safety and health issues.
What are the existing international legal frameworks relevant to child labour in the fisheries and aquaculture sector?	**Part 2/section 4.1** *(International legislation and collaboration)* describes key Conventions and other relevant international instruments. More information is included in **Appendixes 2** and **3**.
What is needed at national level with regard to legal frameworks, institutional structures and tools to effectively address child labour?	**Part 2/section 4.2** *(National implementation and regional collaboration)* points out key considerations with regard to national legislation and institutions. **Part 2/chapter 5** *(Deciding what constitutes child labour)* describes how to proceed when defining child labour in a national context and suggests criteria that can be used for this purpose. **Part 2/chapter 6** *(Closing the data and knowledge gap)* gives guidance on how to make available improved information on child labour. The **Summary of recommendations** provides a specific set of recommendations for governments.

Is it important to engage with organizations of fishers, fish farmers, fish workers and employers, and other sectoral institutions, and their communities, when addressing child labour?	**Part 2/section 4.3** *(Engagement of communities, organizations of fishers, fish farmers, fish workers and employers, and other sectoral institutions)* argues for the importance of multistakeholder participation and the involvement of those directly concerned when identifying practical ways for addressing child labour. This importance is also referred to in many other places in the document, for example, in **Part 2/section 6.5** *(Utilizing information: raising awareness, strengthening capacities and improving policy coherence)* and in **Part 2/chapter 8** *(Closing the data gap)*. The **Summary of recommendations** provides a specific set of recommendations for NGOs and development partners at local level.
Is there a framework for how to take action against child labour?	**Part2/chapter 7** *(Taking action to eliminate child labour)* explains the three pillars of the ILO child labour action framework: prevention, withdrawal and protection.
I am working for a development agency and am planning a new project in the fisheries and aquaculture sector and need to make sure that child labour is adequately addressed. Are there any specific tools I can use?	**Part 2/section 7.2** *(Preventing child labour)* includes a checklist (**Box 15**) to use when engaging in new initiatives to make sure child labour is considered. The **Summary of recommendations** provides a set of recommendations for development partners.
How is education important?	**Part 2/section 7.2** *(Preventing child labour)* includes a discussion on the importance of education.
What can companies, organizations of fishers, fish farmers, fish workers and employers, and other sectoral institutions do?	**Part 2/Section 7.2** *(Preventing child labour)* includes discussions on corporate social responsibility and on technology development. The **Summary of recommendations** provides a specific set of good practices for organizations of fishers, fish farmers, fish workers and employers, and other sectoral institutions.
What is meant by "worst forms of child labour" and how should this be addressed?	In **Part 1/section 1.2** *(Worst forms of child labour)*, the worst forms of child labour are defined and **section 3.3** *(Acceptable and appropriate tasks vs. child labour and hazardous work)* discusses definitions in the context of the fisheries and aquaculture sector. **Part 2/section 7.3** *(Withdrawing children from child labour)* discusses the actions that should be taken.
Are there cases where improved protection may help children?	**Part 2/section 5.1** *(Risk assessments)* describes how personal protective equipment should be one of the last resorts when addressing occupational safety and health (OSH) concerns. **Part 2/section 7.4** *(Protecting children from hazardous work)* discusses protection from hazardous work of children above minimum age for employment.
I need to address child labour (for example, in my project, in the community where I work) but I don't know where to start. Is there any advice on entry points?	**Part 2/chapter 8** *(Closing the data and knowledge gap)* provides strategies for addressing child labour.

PART 1:
CHILD LABOUR IN FISHERIES AND AQUACULTURE: CONCEPTS AND CURRENT SITUATION

The need for knowledge

In order to address child labour, we need to understand what the issues are and how child labour relates to poverty and other surrounding circumstances. Part 1 of this guidance document provides information on child labour, on the fisheries and aquaculture sector – including safety and health issues – and on the overall vulnerability context. The different tasks that children undertake in fisheries and aquaculture are discussed and examples of child labour and hazardous work are identified. The information in Part 1 provides the background to and a framework for the discussion in Part 2 on how to address child labour in fisheries and aquaculture.

Mexico, boy with fishing rod
©FAO

1 What is child labour?

1.1 Defining child labour

According to the 1989 UN Convention on the Rights of the Child,[7] a child is a person under 18 years of age. The 1999 Worst Forms of Child Labour Convention (No. 182) also states that the term "child" shall apply to all persons under 18 years. Not all work performed by children is child labour that must be eliminated. While child labour by definition is unacceptable and should be abolished – in particular the worst forms of child labour, as a matter of urgency[8] – there are age-appropriate tasks carried out by children that are not harmful to them and which can even be beneficial. Especially in the context of family-based economic activities and household chores in a rural environment, some participation of children may be regarded as positive, since it contributes to the intergenerational transfer of skills and children's food security.

Child labour is defined[9] as work that impairs children's well-being or hinders their education, development and future livelihoods. It is work that is damaging to a child's physical, social, mental, psychological or spiritual development, because it is performed at too early an age or is otherwise unsuitable for children (for example, due to the nature or conditions of the work, including the working hours). It deprives children of their childhood, their dignity and their

rights.[10] Accordingly, "child labour concerns work for which the child is either too young – work done below the required minimum age – or work which, because of its detrimental nature or conditions, is altogether considered unsuitable for children and is prohibited"[11] (see **Box 1**).

The Minimum Age Convention, 1973 (No. 138) calls for national policies to eliminate child labour and the establishment of a general minimum working age of 15 (allowing for certain exceptions under specific circumstances, see **Box 2**).

The Work in Fishing Convention, 2007 (No. 188) sets the minimum age for work on fishing vessels at 16 (Article 9(1)). However, the competent authority may lower this to 15 years of age for people who are no longer subject to compulsory schooling but are engaged in vocational training in fishing. People of 15 years of age who are still in school may also be authorized to perform light work during school holidays.[12] The Convention sets a minimum age of 18 for assignment to activities onboard fishing vessels that are likely to jeopardize health, safety or morals. Young persons between 16 and 18 years

---

7 See www2.ohchr.org/english/law/crc.htm for the text of the Convention, usually referred to as the CRC.

8 See Article 1 of Convention No. 182

9 See, for example, Article 32 of the CRC.

10 See the Minimum Age Convention, 1973 (No. 138) and the Worst Forms of Child Labour Convention, 1999 (No. 182). Links to the Conventions are available on the ILO NORMLEX website: www.ilo.org/dyn/normlex/en/.

11 Report of the Secretary-General to the UN General Assembly, dated 27 July 2009 (A/64/172), *Status of the Convention on the Rights of the Child*, Paragraph 13.

12 In accordance with Article 9.(2) of Convention No. 188, in such cases, "it shall determine, after consultation, the kinds of work permitted and shall prescribe the conditions in which such work shall be undertaken and the periods of rest required."

Box 1: What is child labour and hazardous child labour?

A **child** is defined as any person under 18 years of age. **Child labour** is defined on the basis of a child's age, hours and conditions of work, activities performed and the hazards involved. Child labour is work that interferes with compulsory schooling and damages health and personal development. The Minimum Age Convention, 1973 (No. 138) sets the general minimum age for admission to employment or work at 15 years (the Convention allows for certain flexibilities in specific circumstances). For work considered as hazardous, the age is 18. The Worst Forms of Child Labour Convention, 1999 (No. 182) defines worst forms of child labour as all forms of slavery or practices similar to slavery, including the sale and trafficking of children, forced or compulsory labour, such as forced recruitment for armed conflict, the use, procuring or offering of children in commercial sexual exploitation or illicit activities, and hazardous work.

Hazardous work, or hazardous child labour, is thus a category of worst forms of child labour and is defined by Article 3 (d) of the Worst Forms of Child Labour Convention, 1999 (No. 182) as: *(d) work which, by its nature or the circumstances in which it is carried out, is likely to harm the health, safety or morals of children.*

Hazardous work can be described as work undertaken in dangerous or unhealthy conditions that could result in a child being killed, or injured and/or made ill as a consequence of poor safety and health standards and working arrangements. Some injuries or ill health may result in permanent disability. Often health problems caused by working as a child labour may not develop or show up until the child is an adult. Children's engagement in hazardous work is also referred to as hazardous child labour.

The exact types and conditions of work that should be prohibited because they represent hazardous work must be determined in each country by legislation or by the competent authority, after consulting the employers' and workers' organizations. Recommendation No. 190, which complements Convention No. 182, offers guidance on the elements that should be taken into consideration, as follows (Paragraph 3):

(a) work which exposes children to physical, psychological or sexual abuse;

(b) work underground, underwater, at dangerous heights or in confined spaces;

(c) work with dangerous machinery, equipment and tools, or which involves the manual handling or transport of heavy loads;

(d) work in an unhealthy environment which may, for example, expose children to hazardous substances, agents or processes, or to temperatures, noise levels, or vibrations damaging to their health;

(e) work under particularly difficult conditions, such as work for long hours or during the night or work where the child is unreasonably confined to the premises of the employer.

Source: FAO/IFAD/ILO, 2010; ILO/IPEC, www.ilo.org/ipec/.

of age are exceptionally allowed to do such work on condition that their health, safety and morals are fully protected and that they have received adequate specific instruction or vocational training, as well as basic pre-sea safety training. There is also a minimum age of 18 for night work, though the competent authority may again make an exception for training purposes if it has been determined that the work does not have a detrimental impact on health or well-being (Article 9(6)). According to Convention No. 188, decisions on these matters are to be made following consultation with employers'

and workers' organizations (fishing vessel owners' and fishers' organizations, where they exist). Whenever possible, national consultations on child labour issues in the fisheries sector should involve fishing vessel owners' and fishers' organizations, as their in-depth knowledge of the sector can support the decision-making process. Blanket declarations that fishing is hazardous and prohibited to children under 18 years of age are thus avoided, and the emphasis is on providing concrete guidance on specific hazardous tasks and occupations in the sector.

Box 2: Minimum Age Convention, 1973 (No. 138)

The main requirements of the Minimum Age Convention, 1973 (No. 138), include the need for member States to specify a minimum age for admission to employment or work below which no child may be employed. This age should not be lower than the age of completion of compulsory schooling and, in any case, not less than 15 years (or 14 years for a developing country, if this country initially requests this exception at the time of ratification) (Article 2). However, Convention No. 138 leaves many decisions to be taken at national level, including the general minimum age itself; thus, some developing countries have set the minimum age at 16 (for example, Brazil, China and Kenya), while it is 15 in some industrialized countries (for example, Germany, Japan and Switzerland). Convention No. 138 also allows the exceptional permission of light work from 13 years (or 12, where the general minimum age is 14), as long as the work does not interfere with the child's schooling or is physically, mentally or socially damaging (Article 7). In such cases, national legislation must determine the activities that can be considered light work and prescribe the number of hours and conditions in which these light work activities may be undertaken.

For hazardous work, the minimum age must be 18 (Article 3). Some exceptions can be allowed for children between 16 and 18 if specific conditions are guaranteed – that health, safety and morals are fully protected and adequate training and supervision is provided. Convention No. 138 also requires that governments establish laws or regulations to determine the types of hazardous work to be prohibited for persons below 18. All necessary measures should be taken to eliminate child labour (for example, the provision of free and compulsory education to all children up to the minimum age for employment), and penalties for employers who violate the minimum age for employment legislation must be adopted and properly applied.

Convention No. 138 also includes other flexibility clauses: (i) countries may (in so far as necessary) exclude from the application of the Convention limited categories of employment or work, if they list these categories in their first report on the application of the Convention submitted to the ILO and give the reasons for the exclusion (Article 4); (ii) developing countries may initially limit the scope of application of the Convention by specifying, at the time of ratification, the exclusion of some economic sectors (Article 5), although only a small number of States chose to use this flexibility clause, which may not be invoked subsequently to ratification. It is worth noting that compulsory coverage (a range of economic sectors that must be covered even when flexibility is used) includes commercial agriculture (plantations and other agricultural undertakings mainly producing for commercial purposes) except for "family and small-scale holdings producing for local consumption and not regularly employing hired workers" (Article 5, Paragraph 3).

Source: Minimum Age Convention, 1973 (No. 138).

Conventions, international instruments and guidelines relevant to child labour in fisheries and aquaculture are further discussed in **Part 2, section 4.1**, International legislation and collaboration.

1.2 Worst forms of child labour

Worst forms of child labour are defined in Article 3 of the Worst Forms of Child Labour Convention, 1999 (No. 182), and comprise:

a) all forms of slavery or practices similar to slavery, such as the sale and trafficking of children, debt bondage and serfdom and forced or compulsory labour, including forced or compulsory recruitment of children for use in armed conflict;

b) the use, procuring or offering of a child for prostitution, for the production of pornography or for pornographic performances;

c) the use, procuring or offering of a child for illicit activities, in particular for the production and trafficking of drugs as defined in the relevant international treaties;

d) work which, by its nature or the circumstances in which it is carried out, is likely to harm the health, safety or morals of children.

Ghana, fish farmers dividing fish

The last category (d) is what is generally referred to as "hazardous child labour" or "hazardous work".

The ILO estimates that at least 40 per cent of forced labour victims worldwide are children. According to some reports, in certain regions of Africa forced child labour is sometimes "linked to traditional practices of placing children in foster care with relatives in distant cities. While parents are promised education for their children, the boys and girls are often ruthlessly exploited as domestic servants, in agriculture and fishing or in the sex industry" (Andrees, 2008, p. 8).

1.3 Causes of child labour

The main cause of child labour is poverty, which in turn is influenced by the effects of social inequalities, structural unemployment, vulnerability to shocks and demographic and migratory developments. Many children work for their survival and parents depend on their contribution even if they know it is wrong. In other situations, there is lack of awareness, with children working seen as normal because parents do not understand the negative effects and long-term consequences of child labour. Child labour tends to occur in environments with cheap and unorganized labour. In some countries and locations, the demographic structure includes a high child-adult ratio (high dependency rate) and

hence a relatively large number of children (see **Box 4**). Formal education may be of poor quality, limited relevance or difficult to access; combined with high costs of schooling and low levels of parental education, these are additional causes of child labour. In communities in remote rural areas or migrant communities where access to schools is poor, child labour can be expected to be more common. Cultural practices, such as social attitudes towards child labour and children's participation in economic activities and household chores, also contribute to the prevalence of child labour. Gender roles are another factor; many poor parents tend to prefer (if forced to choose) that a boy child attends school while girls are kept at home (to help with household chores[13], childcare and family activities). The absence of appropriate national policies and legislation on child labour and inadequate enforcement of existing legal frameworks further exacerbate the situation.

Table 2 provides examples of factors influencing the supply and demand for child labour. Supply factors (also referred to as "push" factors) refer to the macro level and household situations and decisions that make children available for work, while demand factors ("pull" factors) contribute to creating employment and labour opportunities for children.[14]

1.4 Data on child labour

Official statistics on child labour in accordance with the International Conference of Labour Statisticians (ICLS) standards refer to the 5–17 year age group. There are children under 5 who work, but almost all child labour involves those between 5 and 17 years of age. In 2008, there were a total of 1.586 billion children in this age group in the world.

---

13 "Household chores" refer to work that usually takes place at home and in which children may be involved, sometimes as substitutes for adults carrying out economic activities outside the household. "Domestic work" refers to (formal or informal) employment in someone else's household.

14 See also section 3.4, Links between child labour, poverty and unsustainable fisheries livelihoods.

Box 3: Children and piracy

Piracy, one of the oldest crimes, has re-emerged in recent years and modern piracy activities have been reported in several areas in the world. While there is no systematic collection of data on the issue, child piracy appears to be on the increase. There are parallels between child pirates and child soldiers – children are used by armed groups and criminal gangs because they are vulnerable and can often be manipulated more easily than adults. They also tend to constitute cheap labour and they often lack legal accountability. Fishing boats are sometimes the target of piracy, but pirates are also recruited from fishing communities. Poverty and unemployment are among the root causes of piracy, and children in fishing communities may hence be drawn into these criminal activities.

Source: Whitman et al., 2012.

Box 4: Orphans in sub-Saharan Africa

Globally, an estimated 16.6 million children under the age of 18 have lost one or both parents to AIDS (2009). Fishing communities tend to be particularly vulnerable to HIV/AIDS because of the way fishing and fish trade activities are carried out and organized, including mobility and migratory livelihood strategies.

Close to 90 per cent of these children live in sub-Saharan Africa. Some of the most affected countries are in eastern and southern Africa. Kenya, Nigeria, South Africa, Uganda, the United Republic of Tanzania and Zimbabwe have a total of more than 9 million orphans. While the global number of children orphaned because of AIDS has increased compared to 2005, a positive trend reveals a narrowing of the difference in school attendance between orphans and non-orphans thanks to the global response to the HIV epidemic.

Source: Allison and Seeley, 2004; UNAIDS, 2010.

Of these, 306 million – or 19 per cent – were in some sort of employment. Although some of this work is considered permissible according to Conventions Nos 138 and 182, it is estimated that 215 million children were involved in child labour in 2008 (ILO, 2010a).

According to available statistics, the number of child labourers fell during the last decade – from 246 million in 2000 to a reported 215 million in 2008. Furthermore, the number of those doing hazardous work declined to 115 million in 2008 (a 10 per cent drop compared with 2004). However, this development is not uniform across the world; most progress has been made in Latin America, less in Asia-Pacific, and in sub-Saharan Africa total child labour has increased. Moreover, the trends vary depending on the age group. After a decline at the beginning of the millennium, hazardous child labour in the 15–17 age group increased (in particular for boys) – from 52 million in 2004 to 62 million in 2008 (ILO, 2010a).

Some 60 per cent of the world's child labourers – 129 million children – work in the agriculture sector, including fisheries, aquaculture, livestock and forestry, and child labour is particularly prevalent in the informal economy and as unpaid family labour. It is estimated that only one in five working children is in paid employment; the majority are unpaid family workers (ILO, 2010a). Among those that are paid, many are unfairly paid. There are also self-employed children working to provide for their own basic needs.

Although the ILO Statistical Information and Monitoring Programme on Child Labour (SIMPOC)[15] (established in 1998) has improved the availability of statistics on child labour in general, information on child labour in fisheries and aquaculture continues to be scarce. Global data are not usually disaggregated for fisheries and aquaculture but

15 See www.ilo.org/ipec/ChildlabourstatisticsSIMPOC/lang--en/index.htm.

**TABLE 2: Supply and demand determinants of child labour
in fisheries and aquaculture**

Supply factors ("push" factors)	Demand factors ("pull" factors)

• Prevalence of poverty and need to supplement household income.
• Lack of access to adequate schools and childcare, particularly in remote areas (insufficient number of schools, geographical distance, poor quality and non-relevant curricula).
• Interruption of education or childcare due to migration.
• Inadequate or insufficient information on behalf of parents (for example, perceived irrelevance of education or poor awareness of hazards of certain work).
• Lack of financial services allowing the household to redistribute expenses and income over time.
• Incompatible attitudes, values and norms: children's participation in fisheries and aquaculture considered a way of life and necessary to pass on skills (fishing, net making/repair, fish processing and trading).
• Necessity to cope with shocks such as a natural disaster or the loss of a household breadwinner (accident at sea, HIV/AIDS).
• Children's interest in proving their skills and making a contribution to the family income:
   - Cultural perception of masculinity and desire to earn income, making boys want to go to sea to fish at an early age.
   - Girls wanting to make money work in fish processing and marketing.
• High child-adult ratio (demographic factors).
	• Demand for cheap labour: children are often paid less than adults (or unpaid) and have weaker negotiating power with regard to terms and conditions of work.
	• Insufficient availability of adult labour at peak (fishing) seasons.
	• Need for substitution of adults in household chores and labour when parents are working, sometimes away from home.
	• Demand for special skills and perception that children's fingers are nimble or their (smaller) bodies better for certain tasks, such as net repairs and diving deep distances to hook/unhook the nets from fishing boats.
	• Existence of certain attitudes and perception that children, in particular girls, are more docile workers.
	• Consideration that certain tasks are children's responsibility (for example, feeding fish or fetching water).

Source: Adapted from FAO/IFAD/ILO, 2010; ILO/IPEC-SIMPOC, 2007; ILO, 2002.

are included in agriculture as a whole. Hence, information tends to come from specific case studies and surveys, or may even be anecdotal (Mathew, 2010). Estimates from four developing countries (Bangladesh, El Salvador, Ghana and the Philippines) indicate that child labour in fisheries represents some 2–5 per cent of the total number of child labourers in those countries, and, most strikingly, children (up to 91 per cent of whom were boys) constituted 9–12 per cent of the total fisheries labour force (Allison et al., 2011). However, the limited estimates available indicate large variations both between and within countries. Estimates show that 29 per cent of the total workforce in the fisheries sector in Senegal are children under the age of 15. Children accounted for 27 per cent of crew members and 41 per cent of those engaged in trade-related activities (O'Riordan, 2006). In addition, a survey of child labour on the Baluchistan coast of Pakistan revealed a 30 per cent incidence of child labour, with children accounting for 27 per cent of workers employed in the fishing sector (Hai et al., 2010). According to a survey on Lake Volta in Ghana, at least one child (boy) is employed in the crew on all boats, including small boats holding only two people. However, it is difficult to assess how many of these children should be classified as child labourers and how many are doing tasks

appropriate for their age for a limited time (Zdunnek et al., 2008).

Child labour surveys tend not to take household tasks or chores into account, and the "double-burden" of many children, especially girls – working both at home and in an economic activity – is therefore often overlooked. If non-economic activities, such as household chores, are included in the definition used to calculate child labour, it is clear that more girls work than boys. Girls also tend to work longer hours than boys. It is a major challenge for child labour statistics in general to take into account the hours devoted to household chores and other – often substantial – non-economic activities, as noted in the Resolution concerning statistics of child labour[16] adopted by ICLS in 2008.

The above figures on child labour in fisheries also exclude children in aquaculture activities as estimates are not available. An estimate of the total number of child labourers in fisheries and aquaculture in the world is therefore difficult, but it is likely to be several million.

---

16 See www.ilo.org/global/statistics-and-databases/
standards-and-Good practice guide/resolutions-adopted-
by-international-conferences-of-labour-statisticians/
WCMS_112458/lang--en/index.htm.

SUMMARY POINTS 1

- A child is a person under the age of 18 years.

- A distinction must be made between tasks that are adequate for children, potentially beneficial for acquiring skills and socialization, and child labour. Child labour is by definition harmful and unacceptable and should be abolished.

- Child labour is particularly widespread in the small-scale sector of the informal economy. Poverty is the main cause of child labour, but child labour is highly contextual and its causes (demand and supply factors) and consequences need to be understood in each specific situation. There are also gender-based differences in the causes and consequences of child labour.

- The total number of child labourers in the world was estimated at 215 million in 2008. Data on child labour in fisheries and aquaculture are lacking, but case studies and anecdotal evidence point to significant numbers. Household chores done especially by girls in the household sphere are often overlooked when assessing child labour.

Cote d'Ivoire, girl helping cutting fish
© Nicole Franz

2 The fisheries and aquaculture sector

2.1 Sector characteristics and role in securing livelihoods

Both capture fisheries and aquaculture are extremely diverse sectors, with regard to the techniques used, the environments in which the activities take place and their scale. Activities range from the production and sale of inputs (including fishing gear, bait, aquaculture seeds and feed) and the actual catching, farming and harvesting of fish, to fish processing, marketing and distribution. Production takes place in and around inland and marine waters, but fish marketing and distribution can take fish workers far from the original fish harvesting point. In capture fisheries, a wide variety of fishing techniques are deployed ranging from simple hand-held gear to sophisticated trawls or purse seines operated by industrial fishing vessels. The small-scale sector, which is often informal and based on family labour, employs the vast majority of the world's fishers and fish workers (over 90 per cent of almost 55 million people engaged in the primary sector of capture fisheries and aquaculture); most of those in related activities are small-scale operators.[17] The majority lives in developing countries. With the addition of people involved in research, development and administration linked with the fisheries sector, and assuming that each job in the primary sector creates a number of jobs in auxiliary activities (for example, post-harvest operations and net making) and that each person employed provides for a number of dependents, it is realistic to assume that fisheries and aquaculture ensure the livelihoods of about 660–820 million people, that is about 10–12 per cent of the world population (World Bank, 2012; FAO, 2012).

Fisheries and aquaculture make important contributions to meeting the UN Millennium Development Goal (MDG) on poverty reduction and food security, are a potential source of wealth creation, and provide essential nutritional benefits. Globally, small-scale fisheries and aquaculture contribute more directly to attaining these goals than do industrial-scale operations; nevertheless, the economic contribution of industrial operations may be significant at national level (Béné, Macfadyen and Allison, 2007).

2.2 Challenges and opportunities

Many wild fishery resources around the world are today in a precarious state due to overfishing, irresponsible practices and pollution. In aquaculture, the development and promotion of better management practices have contributed to responsible fish farming but there are still many concerns with regard to environmental

---

17 The small-scale fisheries form a diverse and dynamic sector with characteristics that vary from one location to another. It tends to be strongly anchored in local communities reflecting their traditions and values. Many small-scale fishers and fish workers (employed in associated jobs, in particular in fish processing, distribution and marketing) are self-employed and engaged in both directly providing food for their household and commercial fishing, processing and marketing (FAO, 2010a). For more information on small-scale fisheries, see www.fao.org/fishery/ssf/en).

© ILO / K. Sovannara

Cambodia, two boys fishing

sustainability. While the consequences of resource depletion and environmental degradation may be felt by fishers, fish farmers and fish workers in small and large-scale settings worldwide, the situation in small-scale fishing and fish farming communities in developing countries is often aggravated by poverty and marginalization. Poverty in fishing and fish farming communities is a complex issue encompassing aspects related to social structures and institutional arrangements, including insecure rights to land and fishery resources; inadequate or absent health and educational services and social safety nets; vulnerability to natural disasters and climate change; and exclusion from wider development processes due to weak organizational structures, poor representation and limited participation in decision-making (FAO, 2010b). Migrants within poor fishing communities often belong to the most vulnerable population groups (Njock and Westlund, 2010).

In spite of these challenges, there have been positive developments with regard to the future of

fisheries and aquaculture and related livelihoods. There is increased understanding of the complexity of the poverty and vulnerability context and of the range of coping strategies applied by fishing and fish farming communities to address threats and sustain livelihoods. A human rights-based approach applied to sustainable development can bring together adequate livelihoods and equitable benefits for all members of society – women, men, youth, elderly and children – and sustainable resource utilization through responsible practices (FAO, 2011a).

2.3 General safety and health in fisheries and aquaculture

The fisheries and aquaculture sector is diverse and represents a wide range of activities; likewise, the associated occupational safety and health (OSH) concerns vary depending on the subsector and the circumstances. Fishing from a small boat in an enclosed lake or on a large industrial vessel in the open sea; processing fish in a small-scale traditional way or in a large mechanized factory, or tending to fish in a homestead pond or in an intensive aquaculture system: each system presents different hazards. As men and women tend to have different socioprofessional roles (see **section 3.2**, *Gender considerations*), risks of exposure to these hazards are gender-differentiated.

Fishing onboard vessels

Fishers onboard fishing vessels work under conditions that are quite different from other professional groups. Fishing takes place in a generally dangerous environment and vessels tend to be in constant motion, even in relatively good weather conditions. In bad weather, movements can be violent and unpredictable. There is often no clear separation between working and personal time or space, especially during fishing trips lasting several days. Fishers sometimes live and work on the vessel in cramped and congested conditions, and may be away from home for long periods. Lack of recreational activities, limited access to adequate food and clean water, and fatigue due to long working hours are all problems. Moreover, fishers are often employed under different conditions from other shore-based workers. Many are self-employed or paid in relation to the catch and profit made by their fishing vessel (ILO, 2007; ILO, 2000).

Fishing at sea is probably the most dangerous occupation in the world (ILO, 1999). Many accidents occur as a result of poor judgement during fishing operations related to pressure to increase profits or ensure a decent income. The human factor is estimated to be responsible for 80 per cent of accidents. Increased competition for dwindling fishery resources leads fishers to take bigger risks, or example, working longer shifts, ignoring fatigue, reducing crew sizes, disregarding safety equipment or investment needs and not paying heed to bad weather warnings. On many fishing vessels, especially smaller ones, crews have to work on deck in all weathers, frequently with hatches open, in order to locate, gather and process their catch (FAO,

Oman, boy cleaning fish

© ILO / P. Deloche

2003; Safety for fishermen website[18]). According to a 1999 ILO survey on health and safety issues in the fishing sector, the most common types of accident were stepping on, striking against or being struck by an object, falling and overexertion. The main causes of these accidents were reported to include "rough weather, fatigue, poor technical condition of the vessel, inadequate or inappropriate tools, equipment, personal protective equipment and inattention" (ILO, 2000, box 2.1). **Box 5** presents statistics on fatal fishing accidents from the United States of America.

Fishers also experience other occupational health problems. The 1999 ILO survey indicated that fishers often suffer from skin and respiratory diseases, and consequences of noise and vibration onboard vessels. The survey showed diagnoses including hypertension, coronary heart diseases and cancer of the lungs, bronchus and stomach,

18 See www.safety-for-fishermen.org/en/.

© FAO

Nicaragua, young girl drying fish

fisheries sector, the greater availability of outboard engines, combined with the need to go further offshore to find fish because of depleted inshore resources, increases risks related to engine failure or sudden rough weather for which a small craft may not be suitable. The common characteristics of small-scale fisheries, in particular in remote areas in developing countries – lack of communication and safety equipment, inadequate search and rescue (SAR) services, poor port and landing facilities, and insufficient medical care onshore – exacerbate the risk of accidents and their potential consequences.

Aquaculture

Aquaculture comprises tasks involving a variety of equipment, chemicals, biological agents and physical environments. OSH hazards exist in hatcheries and grow-out facilities (such as ponds, pens and cages) and in feed mills. The aquaculture sector is generally very dynamic and has shown good growth during the last decade; as a result, new OSH hazards may not be well studied (Moreau and Neis, 2009).

Identified OSH risks include musculoskeletal injury due to, for example, heavy lifting or long hours of repetitive hand feeding; physical injuries caused by slips or falls on wet and slippery surfaces; cuts from using knives; and wounds from other equipment or machinery. Chemicals are used for a number of reasons in aquaculture, such as for disease control or to fertilize fish ponds, and include a variety of substances, for example, those associated with structural materials or with

and diseases typically associated with fishers: salt-water boils, allergic reactions to cuttlefish and weeds, fish erysipeloid (a bacterial infection also known as "fish handler's disease"), acute tenosynovitis of the wrist ("fisher's tenosynovitis"), conjunctivitis and poisoning from fish stings (ILO, 2000). Moreover, particularly in tropical inland waters, fishers can be exposed to water-borne diseases (such as bilharzias) or threatened by wild animals (such as hippopotamuses and crocodiles) in lakes and estuaries.

Increased mechanization may have improved to some extent working conditions and efficiency, but many of the underlying issues remain and workers are now exposed to new hazards. In the small-scale

Box 5: Reasons for fatal accidents in the USA commercial fishing fleet

In spite of a gradual decline in fatalities since 1992, a recent survey of commercial fishing deaths in the USA revealed that fishing was still the occupation with the highest number of fatalities in the country. Of the 504 deaths reported in the period 2000–09, most were caused by a vessel disaster, a fall overboard or an injury onboard. The risk factors for vessel disasters varied depending on the type of fishery, but causes included flooding, vessel instability and being struck by a large wave. Severe weather conditions contributed to over 60 per cent of fatal vessel disasters. None of the crew members who died from falling overboard were wearing a life jacket and the majority were alone on deck when they fell. Falls were generally caused by trips or slips, losing balance or gear entanglement.

Source: Centers for Disease Control and Prevention, 2010.

© ICSF

South Africa, women mussel gatherers

soil and water treatments, antibacterial agents, other therapeutants, pesticides, feed additives, anaesthetics, and hormones. Many usages have been adopted from other sectors, in particular agriculture (GESAMP, 1997). Direct contact with chemicals can lead to burns, skin irritation or allergies; inhalation of, for example, formaldehyde, can cause respiratory problems, including asthma; exposure to biological substances (for example, by workers employed in feed milling facilities with poor ventilation systems) can lead to allergies and asthma (Erondu and Anyanwu, 2005).

Risks from handling fish are similar to those in the capture fisheries sector and include cuts, bites and puncture injuries from sharp teeth, spines or bones. In addition, there are other risk hazards when submerging or diving in fish ponds or other grow-out facilities, for example, parasitic infestation

and pathogenic infections (Moreau and Neis, 2009; Erondu and Anyanwu, 2005).

Boatbuilding

Boatbuilders are exposed to a variety of substances and materials (styrene, resins, solvents, paints, welding fumes and coating systems) that can cause injuries, illnesses or allergies. In addition, wood itself can create hazards, and there are toxic types of wood.[19] Inhalation of wood particles, in particular fine wood dust, but also direct contact with hard wood, can lead to poisoning and wood dermatitis. Sawing, planing and sanding wood are activities

---

19 www.wood-database.com/wood-articles/wood-allergies-and-toxicity/.

which can cause asthmas and allergies (Brigham and Landrigan, 1985; Hausen, 1986).

Furthermore, there is the risk of explosions and burns from flammable solvents and other products used in boatbuilding. OSH hazards also include injuries due to falls, repetitive motions and noise (Brigham and Landrigan, 1985).

Fish processing and marketing

Fish and seafood processing involves a wide spectrum of techniques and final products. Large-scale processing can take place onboard factory vessels or at shore-based plants. Freezing is globally the most common method of processing fish, followed by canning. Fish may be degutted and filleted.

In the small-scale or artisanal sector, fish processing often takes place close to the landing site or homestead. Salting and fermenting are common in Asia, and smoking is mainly used in Africa. Throughout the tropics, drying is widely practised. To avoid spoilage, insecticides are sometimes used, including chemical products (such as organocholorine pesticides, applied because less expensive) that do not meet food standard requirements. Such practices entail risks to human health, both for the processor and the consumer.

Skin rashes, allergic reactions and asthmatic symptoms are reported to be common ailments among processing plant workers (Lopata et al., 2005). Health risks also include cuts and injuries from sharp tools. In the artisanal sector, fish smoking can present hazards, for example, in Africa, where many women and girls are involved in smoking fish using inefficient smoking ovens, the dense smoke and heat generated pose a health risk.

Fish marketing may involve the carrying of heavy loads and the manual handling of fish, which can lead to musculoskeletal injuries, allergic reactions and fish erysipeloid. Another aspect of fish marketing that is particularly harmful in small-scale fisheries in developing countries involves potential hazards when transporting fish to distant markets,

where there are risks related to road safety and personal security.

The prevalence of HIV/AIDS in fishing communities is often higher than national averages; women fish processors and traders (and fishermen) may be at risk if transactional sex – fish-for-sex – is practised. Fish-for-sex is a phenomenon that has been observed in many different developing countries but particularly in sub-Saharan Africa. It is an arrangement between female fish traders and fishermen, whereby women secure their supply of fish by making (part of) the payment in sexual services (Allison and Seeley, 2004; FAO, 2007; Béné and Merten, 2008).

Other safety and health considerations

Other characteristics and circumstances that are common to fishing and fish farming communities and which should be considered in the context of safety and health include migration, the high prevalence of HIV/AIDS, gender-based violence, sexual exploitation and drug abuse:

• **Migration** and mobility have a long history in fishing communities and concern not only fishers but also fish processors and traders. Migration used to depend on the movements of the fish, while today there are mainly economic and social reasons to move. Migrating communities often lack secure access to health and other social services and may suffer from marginalization and inadequate integration in local communities (Njock and Westlund, 2010).

• **Prevalence of HIV/AIDS** is often high in fishing communities and is considered to be a consequence of the mobility of fishers and fish workers. Sexual behaviour during frequent travel and migration may be different from the more constrained home norms. Another factor contributing to more risky sexual behaviour is the fact that fishing is a high-risk occupation, and this can lead to risk denial. Fish-for-sex transactions, lack of health services, lack of prevention, treatment and mitigation measures, and daily cash that can be spent on sexual services and alcohol may be other contributing factors (Allison and Seeley, 2004; Westlund, Holvoet and Kébé, 2008).

- **Violence** can result from alcohol abuse – women and girls are particularly vulnerable to sexual abuse. In this respect, the lack of security in some areas is an important concern (ICSF, 2010). Combined with promiscuity at landing and processing sites, where children and youth may spend a large part of both their work and leisure time, lack of security risks translating into early involvement in sex and violence. Smoking and drug use are further potential problems in fishing communities – an ILO-supported study on fishers in Indonesia found

that almost all fishers, even children, smoke, and that alcohol and drug use is acknowledged as a problem (Markkanen, 2005).

Most of the safety and health hazards described here apply to all those who work in the fisheries and aquaculture sector, and should be eliminated, avoided or mitigated. In the context of child labour, it is important to understand that children are at even greater risk than adults. This is because their minds, bodies and judgement are still developing.

SUMMARY POINTS 2

- Fisheries and aquaculture, in particular the small-scale sector, make important contributions to poverty reduction and food security. Over 90 per cent of all fishers and fish workers are small-scale operators and the majority live in developing countries.

- There are still considerable challenges to achieving sustainable livelihoods. In addition to concerns related to overfishing and resource depletion, these challenges include economic and political marginalization, limited access to social services and resources, and often high levels of vulnerability to natural disasters and climate change.

- Fishing at sea is probably the most dangerous occupation in the world. Occupational hazards also exist in the post-harvest sector, aquaculture and upstream activities such as boatbuilding.

- Occupational safety and health concerns can be aggravated by additional factors, such as migration, the high prevalence of HIV/AIDS, gender-based violence and drug use.

Ecuador, family, including young boy, repairing fishing nets
© FAO

3 Children in fishing and aquaculture

3.1 What jobs do children do?

Children engage in a wide variety of activities, from harvesting and farming of fish in capture fishing and aquaculture, to all associated operations (processing, marketing and other post-harvest activities) and upstream industries such as net making and boatbuilding. Children also perform household chores in their fishing and fish farming families and communities. Some of these tasks are appropriate for their age, whereas others should be classified as child labour or even hazardous work (see **section 3.3**, *Acceptable and appropriate tasks vs. child labour and hazardous work*).

Child labour appears to be particularly widespread in the small- and medium-scale sectors of the informal economy. Child labour is related to poverty and is therefore more frequently found in poorer countries and areas. Children often work in small-scale, private or family-based enterprises, frequently as unpaid family labour.

In capture fishing, children are engaged in all phases of a fishing trip, including onshore preparations before leaving, tasks during the trip and work on return to shore or harbour. Activities cover, for example, preparation and loading of gear and other equipment, procurement, carrying and loading of food and water, and launching of the boat. Onboard the vessel, children may engage in a wide range of activities: rowing or steering the

boat; keeping watch; bailing water; casting and pulling nets or using other gear (for example, line fishing); operating machinery; diving to disentangle nets or in other ways attend to gear (reset or check); and sorting and cleaning fish. On return to shore, children help with pulling the boat onshore, unloading, sorting and cleaning the catch, cleaning the net and hull, and boat and net repair. Children may also be involved in near shore collection of fish and shellfish (on foot or by diving), in other types of fishing not onboard vessels (see **Box 6**), or in illicit practices such as fish poisoning and fishing with explosives (FAO, 2010a; Mathew, 2010).

Children also participate in boatbuilding activities, net making and repairing, and other maintenance work, such as boat waxing and painting, upkeep of outboard engines, and general carpentry. The post-harvest sector involves children in a range of activities, from unloading fish from the boat through to transport, marketing and distribution. They perform fish processing tasks such as sorting, peeling, slicing, filleting, salting, smoking, curing, drying and packing. Children help in carrying loads, and transporting and selling fish.

In aquaculture, children usually assist in farm operations, such as feeding and fertilizing, cleaning ponds and harvesting fish. They sometimes collect shrimp or seed (although wild shrimp seed has nowadays been largely replaced by hatchery production), and are engaged in seaweed farming

Gujarat, fisherwomen sorting dry fish

Box 6: *Jermal* fishing in Indonesia

One special type of small-scale fishing found in parts of Asia and which can involve child labour is the use of fishing platforms, called *jermals*. A *jermal* is a platform supported with wooden tree trunks usually built in shallow waters, 7–8 kilometres out at sea. The platforms are built over traps and accommodate men and boys working on the catch who can stay there for extensive periods. The *jermal* catch comprises mostly shrimp and other small seafood.

Source: Markkanen, 2005.

---

and processing, ornamental fish culture, cage culture in rivers and estuaries and mariculture in coastal waters (FAO, 2010; Mathew, 2010).

3.2 Gender considerations

In line with common gender division of labour among adults, boys tend to be involved more in fishing and girls in post-harvest activities. Data on child labour in capture fisheries are limited and information about the aquaculture sector is even scarcer. As with adult labour, gender roles in child labour are variable and should be examined in the local context (see **Box 8**).[20]

In cases of worst forms of child labour, such as trafficking, gender differences may also be noted. Girls appear to be trafficked in particular for commercial sex and domestic labour, boys in particular for work in fishing, agriculture and mining, as well as for armed conflict. In Ghana, in marine and Lake Volta fisheries, girls engage in post-harvest activities (such as sorting, smoking

---

20 For more information on gender in fisheries, see also Dey de Pryck, 2013.

© ILO / M. Crozet

Sri Lanka, fishermen village after tsunami

and selling) in addition to cooking, farm work and running errands. Boys may also cook and do errands, but are more often involved in paddling, pulling nets, diving and carrying loads. In Lake Volta, some girls also go fishing and diving, which further increases their working hours since their participation in fishing expeditions does not exclude them from the traditional post-harvest activities generally reserved for women and girls (ILO, 2007; Afenyadu, 2010).

3.3 Acceptable and appropriate tasks vs. child labour and hazardous work

Not all activities carried out by children are child labour and need to be abolished. Certain non-hazardous light duties can be an acceptable and even beneficial activity for children. Participating in light household chores or activities that are appropriate for their age, with the family or community, can give children an opportunity to develop skills and improve their sense of belonging

Box 7: Example of children's work in inland fisheries in Cambodia

Research in three provinces in Cambodia (Kampong Chhnang, Pursat and Siem Reap) to better understand the extent of child labour in freshwater fishing found that children were engaged in a number of activities related to the making and preparation of boats and gear, fishing and gleaning/ collection of fish products, harvest (ponds or cages) and post-harvest operations. Children start some of these activities well before 12 years of age, and sometimes as young as 4.

Source: MLVT/Winrock, 2011.

© FAO

Malawi, small boy waiting for fish that are being cooked

and marketing skills to their daughters. However, while such arrangements tend to be considered educational, or helping within the family context, the line can be very fine as to whether they are appropriate for children or potentially detrimental and therefore child labour.

Whether an activity is adequate or constitutes child labour, or even hazardous child labour, depends on a range of factors and conditions. Besides the age of the child, there are factors relating to the characteristics of the fisheries and aquaculture professions and the safety and health issues outlined in **chapter 2**, on the fisheries and aquaculture sector. Because of their immature physical and mental state, children are generally more vulnerable to hazardous conditions that already pose threats to the safety and health of adults, for example, bad weather during fishing at sea or continued exposure to smoke during certain types of fish processing. Dangerous and detrimental work should not be carried out by anyone – adult or child – and efforts should be made to ensure that appropriate safety standards and equipment are in place (such as appropriate design of vessels and life jackets), that work conditions improve by adopting better techniques or practices (such as improved fish smoking ovens and training of crew in safety procedures on fishing vessels). Workers exposure to hazards should be limited, including by refraining from going to sea in bad weather or adopting mechanization of dangerous work tasks).

and self-esteem. Children who have reached the minimum age for admission to employment can work full-time in non-hazardous work. Such work becomes an issue when it is hazardous or when children are exploited or under age, and interferes with schooling.

Not all tasks, therefore, are considered child labour. Some activities in fisheries and aquaculture can be positive, providing children with practical and social skills for work as adults. There is often a tradition of fathers passing on their fishing knowledge to their sons, and mothers transmitting fish processing

Some tasks are unsuitable for children precisely because they are children and their bodies are not yet fully developed and hence they are at greater risk than adults (see **Box 9**). These includes

Box 8: Gender roles in fisheries

While the generalization of the professional roles of men as fishers and women as fish processors and sellers is largely correct, a closer examination of gender in fisheries reveals a more complex situation according to local and cultural contexts. In some countries, it is common for women to fish or collect seafood (for example, mussels and clams) in coastal or inland waters. While this is often done as a sideline, it can also be very important for feeding the family. Women may be entrepreneurs as well as fish buyers; it is not unusual for them to advance money to finance fishing trips or give loans to fishers against a guaranteed supply of fish when the catch is landed.

Source: Westlund et al., 2008.

Box 9: Why are children at greater risk than adults from safety and health hazards in the workplace?

Child labourers are susceptible to all the dangers faced by adult workers when placed in the same situation. However, the results of exposure to workplace hazards can often be more devastating and longer-lasting for children, resulting in permanent disabilities or psychological damage as a result of working and living in an environment where they are under physical duress, denigrated, harassed or exposed to violence.

When assessing hazardous work of children, it is important to go beyond the concepts of work hazards and risks as applied to adult workers, expanding them to include the developmental aspects of childhood. Children are still growing and they therefore have special characteristics and needs for example, in terms of their physical, cognitive (thought/learning) and behavioural development and emotional growth. Likewise, their judgement is not fully developed and may lead to additional risk taking.

The consequences of some safety and health problems do not develop, appear or become disabling until the child is an adult; this must be factored in when considering the long-term effects of working as a child labourer. For example, carrying heavy loads as a child can result in long-term musculoskeletal problems later in life, or cancer may develop in adulthood as a result of exposure to chemicals as a child labourer.

For more information on specific developmental differences between adult and child workers, see **Appendix 1**.

Source: ILO, 2006.

Box 10: Examples of worst forms of child labour in fisheries and aquaculture

In the Philippines, children are engaged as swimmers and divers in *muroami* (a type of net) fishing, targeting reef fish – an extremely hazardous form of work. Child labourers are reportedly at risk of ear damage, injuries from falls, shark attacks, snake bites and drowning (ILO, 1998).

Child labourers in shrimp processing (de-heading) depots in Bangladesh tend to work hours that prevent them from attending school. They often work for nine hours without a break in extremely unsanitary conditions, and are frequently cheated of their pay. Cuts to hands and feet are common and can become badly infected, abscessed or swollen. Sexual abuse, including rape, is also reportedly common. For unmarried girls, the very fact that they work in the industry can mean their reputations and marriage prospects are tarnished, regardless of whether or not they engage in sexual activity (EJF, 2003).

On Lake Malawi, young boys are sometimes used for bailing water out of the small fishing boats operating on the lake. These *chimgubidi* ("water pumps") have to work throughout the fishing trip, often all night, and are not allowed to fall asleep or get seasick. If they fail on any of these counts, they receive only half pay, and if they get seasick, they have to drink lake water (to "treat the sickness") (personal communication, fishing community during field trip to Senga Bay, Malawi). On Lake Chilwa, young boys work as "bila boys" to guide and disentangle the seine nets when they are pulled in. This is a dangerous task, because they must be in the water for a prolonged period of time and dive to unsafe depths (Lugano and Zacharias, 2009).

In Ghana, there are reported cases of children being traded as commodities for monetary benefits. They are trafficked through middlemen to distant destinations, unknown to both parents, to work in fisheries, for example, taken from their home villages to catch *kapenta* (*Limnothrissa* spp.) in Lake Volta. The depletion of fishery resources in the lake is ostensibly the reason attributed to this "hiring" of children as workers, as they are source of cheap labour. Their smaller fingers are believed to be efficient at removing *kapenta* from small-meshed gillnets and they often have to dive to release entangled gillnets from tree stumps on the shallow lake bottom. In the process, they are exposed to a high rate of parasitism (for example, bilharzias and guinea worm) and are also at risk of drowning. Night fishing with children leads to high rates of school drop-outs (Sossou and Yogtiba 2009; UNODC, 2006, referred to in Mathew, 2010.)

TABLE 3: Selected list of common fishing and aquaculture tasks, hazards and potential consequences

Tasks	Hazards	Potential health consequences
Capture fisheries		
Sorting, unloading and transporting catches	Heavy loads; large machines with moving parts	Joint and bone deformities; blistered hands and feet; lacerations; back injury; muscle injury; amputation of fingers, toes and limbs; noise-induced hearing loss
Preparing food on fishing vessels	Sharp blades; stoves in poor repair	Cuts; burns
Diving for various aquatic species, to free snagged nets, or to scare fish into nets	Deep water; dangerous fish; boat propellers; fishing nets; entanglement	Death by drowning; hypoxia; decompression illness; dizziness; emphysema; bites or stings from fish; hearing loss from ear infections or rapid pressure change
Active fishing; pulling fish onto boat	Heavy loads; sharp objects	Blistered hands and feet; lacerations; back injury; muscle injury; fish poisoning
Going out to sea	Frequent lack of appropriate fishing ports, boat shelters and anchorages	Death or broken bones from surf crossing
Dangerous fishing operations	Trawling vessel gear snagging on a fastener (because of obstacles on the sea bottom); small seiners capsizing under the downward pressure of a large catch of fish "sinking" during the last stage of net hauling; getting caught up in nets; ropes running out while setting the gear; attacks from marine animals (also for wading fishers)	Death due to capsizing of vessels; sweeping overboard; stings, bites, tail kicks
Working on boats and water in general	Crowded conditions; deep water; cold water; polluted water; slippery walkways; fumes and other odours; loud equipment; lack of drinking water; long hours; working at night; bad weather and lack of poor weather warning systems and radio communication; loss of engine power; fire on board; unsuitable boats (sailing farther offshore on prolonged fishing trips on small fishing crafts built for inshore fishing/day trips)	Death by drowning; hypothermia; nausea; claustrophobia; bilharzias, guinea worm and similar parasitic infections; broken bones and head injuries from slips; physical or emotional abuse; exhaustion; hunger; dehydration; capsizing, grounding, getting lost and collisions (as a result of small boat accidents occurring in sudden gales, major storms and heavy fog)

Long periods at sea on boats or fishing platforms	Sexual abuse, intimidation, exposure to and pressure or enticement to engage in adult behaviours	Sexually transmitted diseases and HIV/AIDS; alcoholism, drug use and smoking-related diseases; diminished sense of self-worth
Behavioural responses to fisheries management	Taking higher risks (if spatial-temporal closures limit the fishing time or area, fishers may venture further offshore)	Death by drowning; physical exhaustion; getting lost
Post-harvest		
Cleaning fish and shellfish; processing, smoking or selling fish	Sharp tools; smoke and chemicals; long hours standing or bending	Blistered hands and feet; lacerations; backache and other musculoskeletal strains and disorders; exhaustion
Repairing nets, vessels	Sharp or heavy tools	Blistered hands and feet; lacerations
Aquaculture		
Tending aquaculture farms	Disease control compounds; mosquitoes	Injury from falls; death by drowning; malaria; dengue; pesticide poisoning

Source: Adapted from IPEC, 2011, p. 26, Table 5.2 and FAO, 2005 (web page), based on Ben-Yami, 2000.

physically demanding work that can cause injuries or harm to children's physical development, or tasks entailing a risk of exposure to noise, poison or toxics (from animals/fish, chemicals or other) or other substances or conditions to which children are likely to have lower tolerance than adults. In fisheries and aquaculture, such activities include operating certain types of gear requiring physical strength, carrying heavy loads (beyond limits specified by national regulations), high-risk diving (at excessive depths or with risk of becoming entangled in gear or of encountering dangerous animals) and handling certain animals or fish.

Duration and timing are factors to consider when determining whether an activity is acceptable for children. Children in the 10–18 year age group require 9.5 hours of sleep in order to support their holistic development, taking into account intellectual, emotional, social, physical, artistic, creative and religious aspects. Night work or long working hours, leading to fatigue, are considered hazardous. In fisheries, working hours are typically irregular and can be extremely long – conditions

which are not suitable for children. Inappropriate tasks include going out on fishing boats at night or attending to and processing fish when landed at very early or late hours of the day.

Table 3 gives examples of tasks, hazards and potential health consequences for children in the fisheries and aquaculture sector (see also **Part 2, section 5.1**, *Risk assessments*).

3.4 Links between child labour, poverty and unsustainable fisheries livelihoods

As in the agriculture sector in general, poverty and social inequalities are the main causes behind child labour in fisheries and aquaculture, but at the same time child labour perpetuates poverty. Child labour has a negative impact on literacy rates and school attendance and limits children's mental and physical health and development, reinforcing poverty and marginalization. Not only is child labour potentially harmful for the child as an individual,

Democratic Republic of the Congo, girl sorting out some dried fish

children are paid less, child labour allows fishing to continue in situations where it would otherwise have stopped because of poor profits. As a consequence, overfishing and unsustainable resource utilization persists reinforcing the vicious cycle.

Poverty and marginalization also work the other way round – the risk that a child engages in child labour is reinforced when conditions of poverty prevail. For example, insufficient access to health care makes the consequences from injuries and other physical and mental problems caused by child labour more severe. This is an issue in many fishing communities located in remote areas or in migratory communities with limited access to adequate social services.

Migration is a common livelihood strategy in many fishing communities and child labour "hotspots" are often linked to situations with high levels of migration. Boys and girls of all ages participate in migration, often already working or training to become fishers or fish workers. Transferring from one place to another tends to negatively impact education, also because receiving areas may have insufficient schools. Moreover, as well as frequent travelling, implicit or explicit demands to help in work may also contribute to children leaving school prematurely (Njock and Westlund, 2010).

there are negative consequences from a broader perspective of poverty eradication, development and sustainable resource utilization. In capture fisheries, using child labour in substitution for adult labour reduce costs – as a result of poor profitability as a consequence of overfishing further exacerbates unsustainable resource utilization. In fact, since

However, the causes and negative consequences and impacts of child labour are highly contextual. In poorer fishing communities, deprived of education and alternative employment opportunities, sons following their fathers into fishing – and girls their mothers into fish processing and marketing – may be

Box 11: Child labour and environmental degradation

There are links between poverty and the environment that can lead to increased child labour. Where fishers start to earn less from fishing, because resources have become scarce or pollution is threatening the ecosystem, the drop in income may force parents to take their children out of school. An illustrative case concerns mussel collectors in North Jakarta, Indonesia. Wives of local fishers often earn a living from collecting green mussels to sell to restaurants, but recently the size of the mussels has decreased and demand has fallen correspondingly. Mussels are sensitive to pollution, and they have become smaller as the environment has deteriorated. In this worsening environment, children face increased pressure to leave school and go fishing with their fathers or engage in other work to help support their families.

Source: ICSF, 2011.

perceived as the only viable options for professional training. If childcare, schools and educational opportunities are inexistent or of low quality, parents may see working and learning a trade as the sole path for their children. In other cases, children may have to work to help their families who cannot afford to send them to school (even when education exists and is free of charge, as there are indirect costs and lost opportunities) (see also **Box 11**). The recent global food and financial crises have been particularly severe for poorer population groups and the effects may be felt for some time yet with increased pressure on households to make ends meet and, as a consequence, on child labour supply. Moreover, if the crisis leads to a cut in national education budgets, it is likely that more children will not attend school and instead become available for work. This is extremely unfortunate as education is key to development and poverty alleviation.

Considering the complexity of child labour and its causes, and its links to the overall poverty context, removing children from child labour in fisheries requires caution, especially if alternatives are scarce, whether with regard to education or to work opportunities outside fisheries (Allison, Béné and Andrew, 2011). While there is no doubt that children's schooling is a priority, a thorough understanding of the precise context is required to tackle child labour in an effective and integrated manner. To achieve sustainable results, it is necessary to address the root causes and deal with specific incidences of child labour. Action is required in education and alternative livelihood arrangements must be developed for improved resource governance and fisheries management. **Part 2** of this document discusses strategies and practices for addressing child labour in fisheries and aquaculture.

SUMMARY POINTS 3

- Children engage in a wide variety of tasks in the fisheries and aquaculture sector, including fishing, pre-trip preparations, post-harvest activities (processing and marketing), feeding and harvesting fish in ponds and cages, boatbuilding and net and making and mending.

- In line with adults' gender roles in fisheries and aquaculture activities, boys tend to be more involved in fishing and girls in post-harvest activities.

- Because of their developmental status, children are more at risk than adults from safety and health hazards. There are many tasks in fisheries and aquaculture that children should not do.

- A priority is the elimination of the worst forms of child labour, that is all forms of slavery, the use of children in prostitution, pornography or illicit activities (such as drug trafficking and piracy) and hazardous work (likely to harm the health, safety or morals of the child). As in other sectors, there is some evidence of child trafficking to work in fisheries.

- Poverty, marginalization and child labour constitute a vicious cycle. Child labour needs to be addressed in an integrated and holistic manner.

PART 2:
GUIDANCE ON HOW TO ADDRESS CHILD LABOUR IN FISHERIES AND AQUACULTURE

The need for action

Although the incidence of child labour globally has decreased by 3 per cent between 2004 and 2008, according to the ILO there is still much to do. Part 2 of this document aims to provide guidance to governments, development partners , organizations of fishers, fish farmers, fish workers and employers, and other sectoral institutions on important considerations and actions to effectively address child labour. It provides an overview of existing international legal and policy frameworks, and emphasizes the need to engage in multi-stakeholder participatory processes with communities to address child labour, in particular in the informal economy. Occupational safety and health (OSH) risk assessments and the Worst Forms of Child Labour Convention, 1999 (No. 182) guide on how to define hazardous work in the fisheries and aquaculture through a specific process. Furthermore, the need for better information, and how to obtain data, is discussed. Part 2 also presents a framework for practical action against child labour, based on strategies defined by the ILO (prevention; withdrawal, referral and rehabilitation; and protection). It provides guidance on entry points, partners and tools and good practices to engage in addressing child labour.

Benin, two boys in a boat
© ILO / P. Deloche

4. Ensuring adequate policy: legal and institutional frameworks

4.1 International legislation and collaboration

The international legal framework for addressing child labour in fisheries and aquaculture consists of a variety of Conventions, other international instruments and guidelines (as mentioned in **chapter 1**). A collection of rights, duties and guidance, the framework provides a policy and legal basis for tackling child labour and offers technical advice. Countries are encouraged to accede to and implement these instruments. However, while some instruments have been widely ratified, translated into national legislation and implemented, others have so far only seen limited applications.

One of the basic documents with regard to children's rights and well-being is the UN Convention on the Rights of the Child (CRC) that entered into force in 1990. It spells out the basic human rights of children everywhere, that is the right to survive, develop to the fullest, be protected from harmful influences, abuse and exploitation, and participate fully in family, cultural and social life. The four core principles of the CRC are non-discrimination; devotion to the best interests of the child; the right to life, survival and development; and respect for the views of the child. The rights spelled out in the CRC are inherent to the human dignity and harmonious development of every child. The CRC protects children's rights by setting standards in health care, education, and

legal, civil and social services; duties lie with others to respect these rights. Article 32 refers specifically to child labour by stating "the right of the child to be protected from economic exploitation and from performing any work that is likely to be hazardous or to interfere with the child's education, or to be harmful to the child's health or physical, mental, spiritual, moral or social development."[21]

The key ILO Conventions[22] relevant to child labour in fisheries and aquaculture were mentioned in **chapter 1**, *What is child labour?*. The Worst Forms of Child Labour Convention, 1999 (No. 182) is one of the most widely ratified Conventions (ratified by 177 countries as of May 2013) and defines worst forms of child labour, including hazardous work. The Minimum Age Convention, 1973 (No. 138) has been ratified by 165 countries (as of May 2013) and sets the minimum age for when children should be allowed to work (see also **Box 2**). Within the overall framework of urging the elimination of child labour, both Conventions assign responsibilities to countries to take action and to consult with employers' and workers' organizations on a number of issues.

---

21 The above description is from the UNICEF website: www.unicef.org/crc/. The full text of the Convention is available from the web page on international law of the Office of the High Commissioner of Human Rights (OHCHR): www2.ohchr.org/english/law/.

22 See the ILO Child Labour web page for full texts of Conventions: www.ilo.org/global/topics/child-labour/lang--en/index.htm.

© FAO

Benin, harbour people meeting

The Working in Fishing Convention, 2007 (No. 188), mentioned in **section 1.1**, stipulates age limits for work onboard fishing vessels.[23] The approach taken by Convention No. 188 is similar to that of Convention No. 138 but is more specific to fishing. The Convention calls to determine, among other things, what activities onboard fishing vessels are likely to jeopardize the health, safety or morals of young people, taking the risks involved into account as well as applicable international standards. The related ILO Recommendation No. 199 provides non-binding guidance on the implementation of Convention No. 188 (see also **Appendix 2**).

In the fishing arena, there are other important international instruments. The FAO Code of Conduct for Responsible Fisheries (1995 – CCRF) is a voluntary instrument that has become probably the most important guiding document for fisheries management and fishing operations globally. It also addresses aquaculture, post-harvest and related activities. The CCRF has no specific provisions for child labour, but covers more generally safety and health standards and adherence to other international instruments. Accordingly, it declares that "States should ensure that safety and health standards are adopted for everyone employed in

fishing operations. Such standards should be not less than the minimum requirements of relevant international agreements on conditions of work and service" (Article 8.1.5).

The FAO 2011 Technical Guidelines on Aquaculture Certification have been established to guide the development, organization and implementation of credible aquaculture certification schemes. They state that "child labour should not be used in a manner inconsistent with ILO Conventions and international standards".

Another development in support of the elimination of child labour in fisheries is the recent agreement of the FAO Committee for Fisheries to develop international guidelines for securing sustainable small-scale fisheries (SSF Guidelines) for which FAO is acting as the Secretariat. Social and gender equality is a central theme, and the guidelines also address child labour in fisheries and aquaculture.[24]

See **Appendix 3** for information on cooperation among UN agencies on fisheries safety issues.

4.2 National implementation and regional collaboration

Ratification and endorsement of international Conventions and agreements are the starting point for action at national level. These commitments need to be translated into national legislation, implemented and enforced. The ILO Conventions give substantial "homework" to countries on precise issues for the elimination of child labour, including identification of worst forms of child labour (Convention No. 182) and setting of minimum ages (Convention No. 138). Both Conventions Nos 138 and 182 cover the process of developing the list of hazardous work, which should possibly include tasks and conditions relevant to fisheries and aquaculture.

In addition, countries need to ensure that the rights of children are protected in accordance with the CRC. Child labour elimination requires

---

23 Convention No. 188 has been ratified by two countries (as of May 2013) and is not yet in force.

24 See www.fao.org/fishery/ssf/en.

legislative coherence in all relevant areas, such as OSH, fisheries management and other sectoral issues; these laws and regulations should be child-sensitive. For example, OSH laws and regulations could restrict the amount of weight that can be carried by children. **Box 12** provides examples of national legislation.

To ensure collaboration and coordination on child labour across different sectors, it is necessary to sensitize officials and establish institutional arrangements between ministries, government agencies and other institutions. Mechanisms and procedures to facilitate collaboration between different actors may be required at both national and local level. They could be in the form of new structures and designations (for example, focal points, child labour units/committees, technical working groups) or additions to the mandates of existing coordination mechanisms (for example, child protection committees, OSH working groups). In areas and situations where migration takes place – a common phenomenon in many small-scale fishing communities – regional and cross-border collaboration is essential, as well as the enforcement of existing international law. Regional organizations, such as Regional Fishery Bodies (RFBs) and Regional Economic Communities (REC), could be useful players in this context – fostering dialogue, harmonizing policies and sharing relevant information.

Laws and regulations only become effective when they are implemented and enforced. Governments are responsible for ensuring that appropriate frameworks and measures are in place while, effective implementation tends to be based on incentives. Incentives can be negative, in the form of penalties for non-compliance, or positive, inducing the desired behaviour. Studies show that a combination of different incentives relevant to and based on an understanding of the national and local context is most effective (Tabatabai, 2003). It may be necessary to strengthen labour inspection, port State control, border controls and other relevant enforcement mechanisms – including training of labour inspectors and other monitoring

Peru, boys eating anchovies

bodies – while ensuring availability of adequate and affordable schools that meet the needs of fishing and aquaculture communities. Economic incentives can involve improved access to markets or financial support to encourage parents to send their children to school (see also **chapter 7**, *Taking action to eliminate child labour*). Special attention is needed where incentives represent a particular challenge in what are often remote small-scale fishing communities.

Institutional and social incentives include "participatory governance arrangements that induce support from stakeholders" and "community-based institutions and social environments that create peer pressure on individuals to comply with agreed-upon community rules" (De Young et al., 2008, p. 83). Such incentives, including buy-in by communities and stakeholders, are particularly important in the informal economy. Community participation has a key role and is discussed in detail below (see **section 4.3** and **chapter 6**, *Closing the data and knowledge gap*).

Box 12: Examples of national child labour legislation: fisheries and aquaculture in national Hazardous Work Lists

National Hazardous Work Lists, a key component of labour legislation required by Convention No. 182, have an important role to play in specifying hazards and occupations relevant for child labour in fisheries and aquaculture. Hazardous Work Lists are not always very detailed, but the examples below provide guidance on how fisheries and aquaculture might be integrated into Hazardous Work Lists.

Côte d'Ivoire

Côte d'Ivoire includes specific reference to work in fishing when citing (Article 11, Arrêté of 2012) prohibited work in the 2012 Hazardous Work List (MEMEASS/CAB, 2012). Specifically, children under 18 years of age should not undertake fishing at sea, in the lagoon or in rivers, nor should they perform deep-sea diving in lagoon and coastal areas or rivers. The List also prohibits more general hazardous work relevant to fishing and aquaculture and which can damage the development of the child, including handling and application of chemical products.

Indonesia

The Hazardous Work List of Indonesia (Decision of the Ministry 2003, Attachment C-5) cites tasks endangering the health and safety of children, including jobs in offshore fishing activities, fishing in deep/pelagic waters and jobs on ships. The List also includes jobs that expose children to general hazardous conditions that can be relevant to fisheries and aquaculture, such as jobs underwater, lifting heavy weights and operating machinery.

Panama

Panama has a very comprehensive Hazardous Work List covering not only underwater work, but also work at high sea and on rivers (including for fishing), and even aquaculture in ponds or lakes (Articles 2, 3 and 7 of the Executive Decree of 2006, Panama). While such a comprehensive list might be considered slightly overprotective, as no child under 18 may work in any of these activities, it demonstrates the legislators' effort to cover in detail otherwise neglected sectors. The List prohibits for children under 18 years of age: jobs underwater (including activities, occupations and tasks that require diving at sea, in rivers, ponds or lakes); jobs at sea, in rivers or other water bodies (including activities, occupations and tasks related to large-scale, small-scale and artisanal fishing, harvesting of shellfish, boat repair and onshore transport); jobs in ponds and fish farms (including activities, occupations and tasks related to aquaculture, such as feeding fish, arranging and repairing nets, pushing boats); and other work in agriculture and fisheries implying specific risks (for example, long hours, hazardous machinery, lifting heavy weights, exposure to toxic substances, organic dusts and extreme temperatures, such as in refrigeration).

4.3 Engagement of communities, organizations of fishers, fish farmers, fish workers and employers, and other sectoral institutions

While appropriate policy, legal and institutional frameworks are fundamental for addressing child labour, such formal arrangements are likely to be insufficient in the informal economy, in particular where poverty is an underlying issue and the traditional organization of work in fisheries and aquaculture includes child labour. Much of children's work involves helping parents or family members, in the vast majority of cases without formal contracts or payment. Such work takes place in the household sphere and includes household chores and fisheries and aquaculture economic activities.

Many communities have limited awareness of the consequences of child labour and – because "children have always worked" – parents and community members may be ignorant about the issue and reluctant to recognize it as a problem. Often, resistance to abiding by rules administered "from above" (central government) and, especially in remote areas, difficulties in monitoring and ensuring enforcement of laws and regulations make them ineffective. While the responsibility of ensuring enforcement of child labour legislation lies with national governments, strong collaboration at grass-roots level is necessary to ensure widespread application. The process of addressing child labour begins with context-specific assessments and cautious debates. The reality of small-scale operations in particular may make it difficult to address child labour; multistakeholder participation, involving children and youth, can help to identify practical and implementable approaches.

Accordingly, governments and development partners need to engage with communities to ensure that parents and communities understand the importance of investing in their children's education and future. This is likely to require not only awareness raising, but also more in-depth training and capacity and organizational development. Such initiatives could be incorporated in existing strategies, plans and interventions by governments or development partners for fisheries and aquaculture management and development, and in communication and extension projects. The required knowledge and competencies must be available within relevant existing community organizational structures, such as school management committees or

fisheries management councils. Collaboration by governments and development partners with relevant local organizations of fishers, fish farmers, fish workers and employers, and other sectoral institutions and NGOs can promote organizational capacity-building. Governments and development partners should also help employers' and workers' associations reach out to the informal economy and self-employed fishers, fish farmers and fish workers. Communities could also be encouraged to establish protection networks and community watch systems to identify, address or refer cases of child labour to the appropriate authorities or organizations for further action (see also **section 6.5**, *Utilizing information: raising awareness, strengthening capacities and improving policy coherence*).

SUMMARY POINTS 4

- There is an international legal framework for tackling child labour, but many instruments still need to be translated into national legislation and implemented. Governments need to ensure that national policy and legal and institutional frameworks are in place to address child labour. However, laws are only effective if they are implemented, and incentives are required to ensure compliance. Regional and cross-border collaboration to address child labour linked to migration.

- In the informal economy, engagement of community together with organizations of fishers, fish farmers, fish workers and employers, and other sectoral institutions is particularly important. Governments and development partners need to engage with stakeholders and partners to ensure community awareness and buy-in are essential.

Guatemala, girl holding crabs
© FAO

5. Deciding what constitutes child labour

5.1 Risk assessments

A critical first step towards eliminating child labour is to understand what constitutes unacceptable and potentially harmful activities for children. A classification can help distinguish between tasks that are appropriate for children, child labour and hazardous work (worst forms of child labour). Risk assessment is an important tool for identifying and addressing safety and health hazards in general. Informed risk assessment, through community participation, incorporating expertise on OSH as well as specific knowledge on fisheries and aquaculture operations, helps determine what types of activity and specific tasks pose risks to children and youth and can help to identify how to eliminate or mitigate these risks.

A risk assessment consists of three main steps: (i) identifying the hazards, (ii) evaluating the risks, and (iii) introducing safety and health measures:

(i). First, identify the hazard(s), defined as the potential to cause harm. Hazards can include bad weather, machinery, tools, transport, processes, substances (for example, chemicals, dust, noise and disease). The aim is to spot hazards that could result in harm to the safety or health of those working. A systematic approach should yield information on:

- what work activities and processes are dangerous;

- how many workers are at risk for each hazardous activity, and whether they are women, men, boys or girls.

(ii). Second, evaluate the nature and level of risk for each hazard identified (different risk reduction measures are required for each hazard). Risk is the likelihood that the harm from a particular hazard is realized.

Box 13: What are hazards and risks?

"Hazard" and "risk" are used frequently in the context of risk assessment and also when talking about OSH and child labour in other respects. A "hazard" is anything with the potential to do harm. A "risk" is the likelihood of potential harm if that hazard is realized. For example, the hazard associated with fishing at sea in bad weather could be falling overboard and drowning. The risk is high if the vessel is not built and equipped to the required safety standards for the scheduled operations and the expected weather conditions. If, however, the vessel is fitted with safety equipment and proper safety procedures are used, the risk is likely to be lower.

Source: Adapted from IPEC, 2011.

TABLE 4:   The components of an OSH risk assessment

Safety and health risk assessment form		
STEP 1 **IDENTIFY WORK HAZARDS** and possible **INJURIES or HEALTH EFFECTS** (for each hazard)	**STEP 2** **EVALUATE RISKS** (for each hazard)	**STEP 3** **INTRODUCE SAFETY AND HEALTH MEASURES** (for each hazard, use the list below in order)
List the main hazards per work activity, plus possible safety or health problems resulting from the hazard(s)	• Adult female worker • Adult male worker • Young female worker • Young male worker	1. Elimination of risk 2. Substitution of risk 3. Technology measures 4. Work organization, information and training 5. Medical/health control measures 6. Personal protective equipment

(iii). Finally, identify the safety and health measures to implement for each hazard in order to prevent or reduce the risk of occurrence of fatalities, injury or illness. There are six categories of measures to be applied in ascending order; a category 2 measure should only be considered when a category 1 measure is not feasible or is insufficient; a category 3 measure is considered only when category 2 does not work and so on. Children need to be removed from hazardous tasks or occupations if the risks are considerable and cannot be eliminated. The six possible categories of measures are listed below.

1. **Eliminating the risk** – always the best solution. For example, do not allow fishing in certain kinds of weather.

2. **Substituting technology to reduce risk** – the next best risk reduction option. For example, substitute a toxic chemical used in aquaculture with a less toxic one.

3. **Introducing new or additional technology** – potentially effective to reduce the risk if elimination and substitution are not

feasible. For example, soundproof a noisy machine, install dust-extracting equipment in the boatbuilding workshop, or use a wheelbarrow or hand cart to carry heavy loads of fish or nets.

4. **Using safe work practices, procedures and methods, linked to appropriate information and training** – appropriate for specific activities to ensure that tasks are carried out in a safe – or safer – manner. Such measures require good organization of the workplace and OSH training for workers. The Work in Fishing Convention, 2007 (No. 188) specifically stipulates countries' obligation to develop regulations obliging vessel owners to establish "on-board procedures for the prevention of occupational accidents, injuries and diseases, taking into account the specific hazards and risks on the fishing vessel concerned" (Article 32).

5. **Providing preventive health measures** – to help workers not fall ill by detecting potential harm or warning signs early on when carrying out hazardous work. For example,

introduce regular lung function tests for workers exposed to potentially harmful levels of dust.

6. **Promoting utilization of Personal Protective Equipment (PPE)** – never the first way to protect workers, but the last resort. While PPE to protect from hazardous substances (such as pesticides) does not guarantee safe use of chemicals, and therefore does not change the nature of hazardous work, other types of PPE (such as specific items of clothing) can drastically reduce the risk of hazard occurrence. For example, in fishing, especially onboard vessels, the availability and use of personal flotation devices or selfinflating life jackets constitute a basic requirement that dramatically reduces the risk of drowning in the event of the wearer being washed overboard. PPE should be provided in addition to other safety and health measures; it must be of good quality to provide genuine safety and health protection. Children's PPE need to be the right size to provide protection against drowning; inspections should be carried out to ensure that PPE is onboard and in working order. It is also important to provide training on the proper use and maintenance of PPE.

© ILO / M. Crozet

Port of Genoa, sign indicating the safety regulations to respect in ship construction site

Guidance on risk assessments on fishing vessels, is available in binding and voluntary instruments prepared by the ILO, IMO (International Maritime Organization) and FAO, and referred to in this document. In accordance with the Work in Fishing Convention, 2007 (No. 188), government authorities should establish the necessary framework to ensure that "fishing vessel owners, skippers, fishers and other relevant people be provided with sufficient and suitable guidance, training material, or other appropriate information on how to evaluate and manage risks to safety and health onboard fishing vessels" (Article 32). Moreover, on the issue of medical health checks, Convention No. 188 stipulates that, in general and in particular for workers onboard vessels 24 metres in length that normally remain at sea for more than three days, no fishers shall work "without a valid medical certificate attesting to fitness to perform their duties" (Article 10).

In the context of child labour, risk assessments should focus on risks to children, with specific attention to gender-based differences in the risk of exposure to hazards, and to vulnerable groups (for example, young migrants, indigenous people and ethnic minorities). as noted in **section 3.3** (see **Box 9**), children tend to be at greater risk than adults. This means that hazards identified for adults need to be further evaluated in relation to children, using appropriate criteria and standards. It also means that work tasks that are not considered potentially harmful for adults may be hazardous for children and should hence be included in the risk assessment. Generic OSH risk assessment has already been done, it could constitute a useful starting point; however, further assessment with

© ILO / P. Deloche

Viet Nam, boy opening scallops

a specific focus on evaluating the risks posed to children is necessary to identify hazardous work for children. IPEC in Cambodia has developed special checklists for self-monitoring of OSH for children in different types of fisheries-related workplaces. These checklists provide an easy system for assessing compliance and awareness of child labour regulations, identifying hazards, risks and safety measures taken in the workplace, and checking availability of personal welfare facilities (such as safe drinking water). Once the checklist has been filled in, it can be used for taking action on priority issues (IPEC-ILO Cambodia, 2003[25]).

Table 4 summarizes the different steps and components of a risk assessment.

The process to develop national lists on hazardous work is discussed in **section 5.2**. This is linked closely to the first part of the risk assessment process described above (identification of hazards). **Chapter 6** then proposes actions for improving availability of data on child labour, which is particularly important in the second part of the risk assessment process (evaluation of the nature and

25 Available at: www.ilo.org/ipecinfo/product/download.do;jses
sionid=df0d8b7cc01a4d3477936d02bf23c0cd55a7c5ae3
8959d2258a8ab7e53c5796e.e3aTbhuLbNmSe3uKci0?typ
e=document&id=14655.

level of risks). **Chapter 7** addresses how to eliminate child labour, including risk reduction measures – as covered by risk assessment step 3 (Introducing safety and health measures).

5.2 Hazardous work lists

Risk assessments are a potential tool for governments in the process of identifying hazardous work for children in accordance with the Worst Forms of Child Labour Convention, 1999 (No. 182). Convention No. 182 requires countries to draw up a list of hazardous work activities and sectors that are prohibited for children in consultation with workers' and employers' representatives (Article 4). Moreover, the Work in Fishing Convention, 2007 (No. 188) requires countries to determine what activities should and should not be carried out by children under 18 years of age, and by children between 15 or 16 and 18 of age (Article 9). Countries must also identify where hazardous work is found and devise measures to implement prohibitions or restrictions based on the hazards included in their list. Because hazardous work lists are critical to subsequent efforts to eliminate hazardous child labour, Convention No. 182 emphasizes the importance of a proper consultative process, especially with workers' and employers' organizations, for the drawing up, implementation and periodical revision of the list. Additional consultations with relevant socioprofessional associations in the fisheries and aquaculture sector, including fishers, fish farmers, and fish processors' and traders' organizations can constribute relevant expertise.

Advice for governments and social partners on what constitutes hazardous child labour is given in Paragraph 3 of the Worst Forms of Child Labour Recommendation, 1999 (No. 190), which accompanies Convention No. 182: "In determining the types of work referred to under Article 3 (d) of the Convention, and in identifying where they exist, consideration should be given, inter alia, to:

(a) work which exposes children to physical, psychological or sexual abuse;

(b) work underground, under water, at dangerous heights or in confined spaces;

(c) work with dangerous machinery, equipment and tools, or which involves the manual handling or transport of heavy loads;

(d) work in an unhealthy environment which may, for example, expose children to hazardous substances, agents or processes, or to temperatures, noise levels, or vibrations damaging to their health;

(e) work under particularly difficult conditions such as work for long hours or during the night or work where the child is unreasonably confined to the premises of the employer."

5.3 Criteria relevant to fisheries and aquaculture

Based on the framework and good practices described in **chapter 4** and **sections 5.1** and **5.2**, government, fish and fish farmer organizations, workers' and employers' organizations, and development partners should collaborate to identify critical activities, circumstances, substances and processes that can represent hazardous work for children in the fisheries and aquaculture sector. Governments should then establish criteria for evaluating hazards in the sector. These need to be appropriately defined in the national and local context and should facilitate the drawing up of national and/or local lists of hazardous work to support the establishment of appropriate national legislation on child labour in fisheries and aquaculture.

Particularly in situations with limited resources, the focus should be on types of work in which children are most commonly engaged. Priority hazards may vary depending on the types of fisheries and aquaculture activities that exist or are most widespread in a particular location or country.

Examples of criteria for the definition of hazardous work in fisheries and aquaculture include:

India, boy loading ice from a truck

In capture fishing:

- The hours required to be onboard the vessel at sea. Children should not be out at sea at night and the number of hours per day and week should be limited (see the ILO Work in Fishing Convention).

- Weather conditions and distance from shore in combination with the size and type of vessel and availability of safety equipment as well as SAR services. Children should not be exposed to sun or cold over many hours, nor to extreme temperatures.

- The type of gear used and the physical strength required. Hazardous mechanized work processes should not involve children.

- If diving, the depth required and the potential hazards in the form of gear entanglement or exposure to animals or plants that can cause harm.

© FAO

Malawi, boys pulling net

- The working (and living) conditions onboard the vessel with regard to space, availability of sanitary facilities, and food and drinking water.

In boatbuilding:

- The level of exposure to noise, dust, sawdust and toxic chemicals.

- The need to operate tools or be involved in work processes that may be dangerous.

- In aquaculture:

- Exposure to toxic chemicals, including fertilizers.

- The potential risk of exposure to water-borne diseases (in freshwater) or other risks (such as gear entanglement or dangerous animals or plants) when submerged in water or diving.

- The existence of security risks, especially if working at night or carrying out pond-guarding functions.

In fish processing and marketing:

- Exposure to potentially hazardous substances, such as smoke (in artisanal fish processing) or chemicals such as insecticides.

- The need to use tools (such as sharp knives) or be involved in work processes that could be

considered dangerous (such as in larger-scale processing plants).

- The requirement to travel for long hours, or at night, including the existence of security risks.

- The perceived necessity to use potentially illegal, immoral or otherwise harmful practices such as fish-for-sex transactions.

- The need to carry heavy loads.

In all subsectors, it is important to include – in addition to those types of hazards mentioned above from the Worst Forms of Child Labour Convention, 1999 (No. 182) – more general considerations such as the number of working hours and to what extent work interferes with schooling. Work that stops a child from going to school or that interferes with schooling and school performance, is considered child labour. This is very relevant to capture fishing since going out to sea (or on a lake) is often unpredictable: it is not always possible to know how long the fishing trip will last.

Consideration should also be given to household chores, which can include physically hazardous tasks (for example, carrying heavy loads of water or firewood), mentally demanding activities (for example, taking responsibility for younger siblings) or long hours that prevent the child from going to school or getting enough time to play or even sleep. This is especially important for children, mostly girls, who combine work in economic activities with household chores.

5.4 Small-scale operations in the informal economy

The basic steps in a risk assessment and in deciding what is hazardous work for children are the same whether carried out for a large-scale workplace (a fish processing plant or larger aquaculture farm), a small family business (a small-scale fishing boat or fish smoking facility) or a sector or industry segment (a certain type of fishery or aquaculture). The scale and scope vary, however – for example, addressing hazardous work in small-scale operations is more complex because poverty is often an issue. To effectively eliminate child labour in a poor fishing or fish farming community, alternative and flexible approaches are required along the lines already described in **section 4.3**, *Engagement of communities*. The following recommendations are given to government and development partners:

- Engage with professional organizations, fishers' and fish farmers' organizations, workers' and

employers' organizations, local NGOs and other community organizations.

- Assess child labour in a wider context, gaining understanding of how it is linked to poverty, social injustices, gender roles, youth employment opportunities and the availability of education and social services.

- Ensure multistakeholder participation in assessments and action planning, holding separate discussions with different stakeholder groups (men and women) where necessary to allow everybody to voice their opinions.

- Take account of the views of children and young people themselves and create awareness of their rights.

- Address child labour within the given context and recognize that abolishing child labour is a long-term process and commitment requiring a multisectoral approach.

Entry points, partners and tools are also discussed in **section 8.1** and recommendations for different stakeholder groups – including governments and development partners – are given in the final part, **Summary of recommendations**.

SUMMARY POINTS 5

- When deciding what constitutes child labour and drawing up hazardous work lists in accordance with the Worst Form of Child Labour Convention, 1999 (No. 182), governments, workers' and employers' organizations, and development partners can use risk assessments to define hazards, evaluate risks, and identify measures to eliminate or mitigate risks.

- Criteria for defining hazardous work onboard a fishing vessel could include hours at sea, weather conditions, type of gear used and related work processes, need for diving, and general working (and living) conditions onboard the vessel. The post-harvest sector, boatbuilding and aquaculture entail other potential hazards, including exposure to smoke (when smoking fish), noise (in a boatbuilding workshop) and use of toxic substances (in aquaculture).

- Consideration should also be given to household chores, which can be physically hazardous and mentally demanding.

- The reality of small-scale operations in the informal economy is complex. Governments and development partners need to use appropriate and alternative approaches to find feasible solutions for addressing child labour in each specific context.

Lake Malawi, happy children in the beach
© Nicole Franz

6. Closing the data and knowledge gap

6.1 Why is information needed?

While there is a general understanding of the causes and consequences of child labour, including its link to poverty (as described in **Part 1**), there is still much to learn about these relationships (especially in specific local contexts) in order to effectively address child labour. Given the complexity of child labour and the need to engage with communities, it is crucial to understand the circumstances in which it takes place. This applies to decision- and policy-makers, development partners and communities themselves. Information and data are also needed to fulfil international obligations (for example, to develop hazardous work lists and further assess the risks for children involved in work in fisheries and aquaculture) and update national policy and legislative frameworks. Closing the knowledge gap implies both improving the availability of quantitative and qualitative data and enhancing access and use of available information.

6.2 Improving data collection and analysis

There are two main strategies for achieving better data availability. One is to integrate child labour into existing data collection and information procedures and systems. The other is to conduct specific risk assessments and surveys explicitly addressing OSH, hazardous work and child labour

in a specific location or subsector. A combination of the two strategies can prove most useful to obtain sufficient data for longer term statistical analyses and monitoring, and for planning and implementing specific interventions.

A wide range of quantitative and qualitative information may be required to understand the specific causes behind child labour and whom it affects, and to identify solutions. Quantitative data on child labour should be sex-disaggregated; qualitative information on work and education should include the views of children and youth, male and female. The very process by which information is obtained can play an important part in addressing child labour by allowing for participatory assessments, discussions and awareness raising.

6.3 Integration into existing information systems

Governments need to make an inventory of existing information sources, data collection exercises, and national statistics and survey instruments, and identify where to include child labour data collection. This is likely to be the most cost-effective way of improving basic information on the subject. Some specific good practices in this respect include:

- Adapt and integrate aspects of child labour in fisheries and aquaculture into standard household and living standard measurement

© ILO / M. Crozet

Sri Lanka, post-tsunami survey

surveys (LSMS) through the introduction of sector modules or specific questions, in addition to oversampling to obtain representative information on specific hotspots.

- Ensure sufficient data disaggregation in relevant surveys: all data should be disaggregated by age (taking account of thresholds of the Minimum Age for Employment Convention and national age limits) and by sex and profession/occupation; details should be provided relative to specific activities and times.

- Compile, compare and extrapolate data and results from existing surveys, at both national and local level, for example, population and agriculture censuses, LSMS and assessments carried out for SIMPOC[26] purposes.

- Compile existing data and information on occupational injuries and diseases to improve the understanding of occupational hazards, for adults and children.

- Seek innovative solutions and proxy variables, and combine different sources of information – including indigenous knowledge and the understanding of the environment and management of fisheries and aquaculture resources – to overcome the limitations of the under-reporting of sensitive child labour information.

6.4 Specific assessments

In line with the rationale of the risk assessment process and in particular if supported by appropriate OSH legislation, businesses and employers in the formal sector can be requested by their governments to carry out surveys in the workplace under their responsibility to identify hazards, risks and security measures in relation to children. However, while risk assessment and reporting obligations as an employer's responsibility can more easily be undertaken by businesses and industries of a certain size, it is less feasible in the

26 See section 1.4.

small-scale, family-based and informal economy. Community-level organizations and NGOs could play a role in implementing risk assessments to identify hazards and viable alternatives. In general terms, it is advisable to develop a culture of OSH for all, promoting adequate and safer technology and raising awareness of the costs of not investing in OSH in terms of, for example, health expenses, loss of human capital and diminished productivity.

Action-oriented research and case studies can help investigate specific dimensions. An example is the work on Lake Volta by the Centre for Advanced Training in Rural Development and FAO. The methodology used in this work is briefly described in **Box 14**.

Risk assessments and other surveys and research conducted by public authorities, development partners, research institutes or other stakeholder should always be carried out in collaboration with those working in the sector, through workers', employers' and fishers' and fish farmers' organizations, and other relevant stakeholders. Workers, including children, and their organizations should be asked for their views on the dangers of the job(s) they carry out. Consideration should be

Cambodia, Survey on child labour in fisheries and aquaculture

© NIS Cambodia / Heang Kanol

Box 14: Study on child labour in agriculture in Ghana

The research on child labour in fisheries by the project "Child Labour and Children's Economic Activities in Agriculture in Ghana" was based around case studies in two districts of Volta Lake and two districts in coastal areas. Several different methods and data collection strategies were used, for example:

- Interviews with key informants/experts (such as representatives of different ministries and government institutions, international organizations, members of district assemblies, trade unions, NGOs and traditional authorities)
- Focus group discussions with community members (such as participants in Child Labour Committees, teachers and associations of fishers and fish processors)
- Semi-structured interviews with children, parents, employers and teachers
- Observations at places where children work (such as landing or processing sites).

Interview results from different sources were triangulated to cross-check the validity and credibility of information. Working children were interviewed as well as their employers, and guardians or parents. Contacts were made through staff of the District Assemblies, NGOs and traditional authorities. The aim was to cover typical work situations and conditions of children in fishing and to explore factors contributing to situations of child labour. Furthermore, the mixture of methods and stakeholders consulted generated diverse knowledge on possible entry points for reducing hazardous forms of child labour and allowed to explore varied perceptions of accessibility, quality and relevance of education. As far as possible, children's views and opinions on work situations were assessed through, for example, the recalling of their activities of the previous day and biographic interviews. A set of specific core questions was used for different stakeholder categories, in order to capture specific perspectives and knowledge.

Source: Zdunnek et al., 2008.

© ILO / M. Crozet

Democratic Republic of the Congo, former child soldiers, pirogue and fishing nets received at the end of a training, in Lake Kivu

6.5 Utilizing information: raising awareness, strengthening capacities and improving policy coherence

Improving the information available and working closely with stakeholders on child labour issues are two important strategies for raising awareness of child labour, its causes and likely consequences. Enhanced awareness and capacity development what is at stake and what can be done, are needed at local and national level, among communities (including children and youth), governments and their development partners, socioprofessional organizations (such as fishers' associations and other producers' organizations), and employers' and workers' associations. Awareness campaigns should target media, schools, parents and guardians, as well as all those involved in income generation (through, for example, skills and entrepreneurship training and microfinance schemes) and fisheries management and development programmes. At community level, it is essential to secure community engagement through participatory assessments and actions participatory.

given to the potentially different situations, needs and perceptions of men and women, boys and girls. Lessons can be learned from experiences of previous accidents and work-related ill health. Close cooperation with those involved and advice from appropriate experts are necessary to ensure correct analyses and to promote the future implementation of the measures proposed on the basis of the assessment.

Governments and development partners to be able to support communities and actors at local level in their efforts to eliminate child labour – and take action in child labour legislation at national level. Knowledge and capacity are required not only at the level of the ministry directly in charge of child labour issues, but also among those working in fisheries and aquaculture administrations and organizations, at national and local level. This cross-sectoral capacity development is important for policy coherence. Child labour considerations should be integrated into fisheries and aquaculture frameworks, for example, when establishing fisheries management plans and community development strategies. Likewise, fisheries and aquaculture aspects should be included in child labour policies and strategies. Integrating and mainstreaming fisheries and aquaculture child labour concerns in relevant strategies and programmes (in a way similar to gender mainstreaming) is likely to achieve the most benefits. **Box 15** provides a checklist that can be

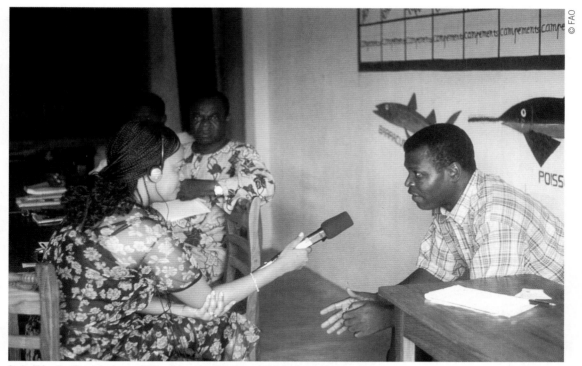

© FAO

Benin, FAO radio officer interviews an NGO participant in a fisheries project in the town port

used for this purpose and **chapter 7**, *Taking action to eliminate child labour*, discusses strategies and activities for addressing child labour. Information and knowledge are important inputs in any action programme and will guide targeting, prioritization and the selection of approaches.

SUMMARY POINTS 6

- More quantitative and qualitative information on child labour is needed to understand its causes and consequences. Integrating data collection needs into existing information systems and processes, and by carrying out specific assessments in collaboration with stakeholders can generate interesting data.

- Information feeds into awareness raising at all levels. It is important for cross-sectoral capacity development in support of policy coherence; child labour concerns should be taken into account in fisheries and aquaculture policies and programmes, while the characteristics of fisheries and aquaculture need to be considered in child labour strategies.

Barbados, children learning about fish
© Nicole Franz

7. Taking action to eliminate child labour

7.1   Action framework

The areas discussed above (in **chapter 4**, *Ensuring adequate policy, legal and institutional frameworks,* **chapter 5**, *Deciding what constitutes child labour,* and **chapter 6**, *Closing the data and knowledge gap*) provide a basis for addressing child labour. The considerations and actions proposed should be integral parts of holistic and participatory approaches that also contain concrete actions to eliminate child labour.

Different types of action are needed. The ILO groups action for the elimination of child labour into three main categories:

- **Prevention** is the primary long-term aim. It means identifying children at potential risk, and stopping them from becoming child labourers in the first place by keeping them out of unsuitable work, especially hazardous work labour, and ensuring adequate alternatives, especially schools.

- **Withdrawal, with subsequent referral and rehabilitation** of children from child labour is immediate action aimed at physically removing children from the situation of child labour, and it is often accompanied by complementary interventions such as referral.

- **Protection** is targeted at children who have achieved the minimum legal employment age (15–17 years depending on the country)

but who are (or risk becoming) engaged in hazardous work. Protecting them from child labour includes improving OSH, working conditions and workplace arrangements.

Different actions for preventing child labour, and withdrawing and protecting children from child labour are discussed below.

7.2   Preventing child labour

Addressing poverty

Investment in the prevention of child labour is the most cost-effective approach to ending child labour in the long run and should therefore be the primary long-term strategy. It means tackling the root causes of child labour so that children at potential risk never become child labourers in the first place. By addressing poverty and promoting inclusive and sustainable development, children stand a better chance of keeping out of child labour and especially hazardous work. To ensure that parents see schooling as the best option for their children, families need, among other things, income security and social benefits (for example, health insurance and safety nets) to survive short- and long-term crises. Access to good quality and relevant schooling and childcare is another key factor; education has to be seen as a worthwhile investment in the future.

© FAO

Ethiopia, school children eating lunch

In many small-scale fishing and fish farming communities, poverty is a complex issue, and as such requires an integrated approach.[27] In many cases, there is a need to improve fisheries management and promote responsible fisheries, including through better participation of organizations of fishers, fish farmers, fish workers and employers, and other sectoral institutions. Small-scale fishers and fish workers and their communities need to have secure rights to the fishery resources that they depend on and to land in coastal areas. However, the issue of resource sustainability is not always the main concern of small-scale fishing communities; they may face more pressing problems in the form of daily needs, ill-health and lack of social services. Therefore, efforts to bring about responsible fishing and sustainable fish farming need to be combined with social and economic development in order to create the incentives and ability to engage in fishery

resource management. By applying a rights-based approach to fisheries management, it is possible to address the broader human rights aspects of fishing community livelihoods, while securing sustainable resource utilization. A holistic livelihood approach working across sectors and and engaging different stakeholders is most effective. For example, considering micro-insurance schemes, or investments in health and education. Education is particularly important for eliminating child labour and is discussed in more detail below.

As mentioned in the context of policy coherence (see **section 6.5**, *Utilizing information: raising awareness, strengthening capacities and improving policy coherence*), programmes should entail cross-integration of child labour and fisheries and aquaculture considerations and aspects. Programmes working towards the elimination of child labour in fishing and fish farming communities need to take into consideration the sector's specific characteristics; child labour should then be a

27 See also chapter 2, The fisheries and aquaculture sector.

Box 15: Checklist for addressing child labour in fisheries and aquaculture support programmes

When planning a support programme in fisheries and aquaculture, in particular for small-scale operations in the informal economy, this simple checklist of questions may help capture child labour considerations. The checklist, developed for local level action, can be adapted to national level programmes.

Policy, legal and institutional frameworks
- What government agency is responsible for coordinating work on child labour? What other relevant institutional structures and organizations are there? Are there organizations (such as NGOs) that work on child rights? What coordination mechanisms are in place?
- Do government policies and legislation exist for child labour? Are they implemented and enforced?
- Are child labour and hazardous work in fisheries and aquaculture defined? What is the minimum legal age for employment? Is light work defined by regulations? Is there a list of hazardous work in accordance with the ILO Conventions? Does it cover fisheries and aquaculture? Have OSH risk assessments been carried out that are of relevance to child labour in fisheries and aquaculture?
- Are there any (local or national) good practices on child labour elimination that can be applied?
- What is the level of awareness on child labour issues among (local and national) decision-makers?

Occupations, working conditions and child labour
- What fisheries subsectors are present (capture fishing, fish farming, fish processing and marketing)?
- What techniques and production systems are used (for example, fishing techniques, types of vessels/craft, intensive/extensive aquaculture, species produced and types of production systems, fish processing methods, marketing and distribution channels)?
- Where is fish marketed and how is it transported?
- What other auxiliary activities such as boatbuilding and net making exist?
- In which of the above-identified activities are children working? What tasks and jobs do they do? Do the same children also work outside fisheries and aquaculture (doing what?) or do they tend to stay within the sector?
- Is there hazardous work in the sector and what processes, equipment and work tasks and conditions are considered hazardous for children and/or for adults? Why and how? (see also step 1 of risk assessment in **section 5.1** and criteria relevant to fisheries and aquaculture in **section 5.3**).
- How many children (and of what ages, boys or girls) are involved in child labour and/or hazardous work? (see also step 2 of risk assessment in **section 5.1**).
- How many hours do children work and what do they do? Does their work interfere with schooling?
- What are the main reasons for children working? Are there different reasons for boys and girls, or for different age groups?

Education and other aspects of community life
- Is school compulsory, available and affordable? To what age and grade is school compulsory?
- Are there any incentives in place for attending school (such as school feeding programmes)?
- Are school curricula and school hours suitable for children working in fishing and aquaculture?
- Do youth easily find adequate employment and decent work? Is there unemployment and who is most likely to be unemployed (young men/young women)? Why?
- What are (local) government plans for schooling and education facilities in the future?
- If children do not go to school, what are the reasons?
- What social services are available in the area/community? What are the main perceived poverty and vulnerability factors?
- Are there functioning community and socioprofessional organizations, including organizations of fishers, fish farmers, fish workers and employers, and other sectoral institutions, and community fisheries management organizations?
- Have there been or are there currently any actions against child labour at local level? By whom and on what scale?
- What is the level of awareness of child labour issues among community members?

FIGURE 1: The virtuous cycle of appropriate actions

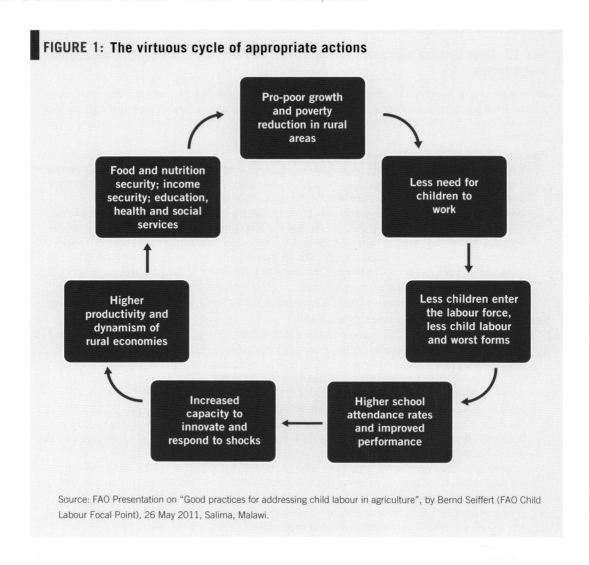

Source: FAO Presentation on "Good practices for addressing child labour in agriculture", by Bernd Seiffert (FAO Child Labour Focal Point), 26 May 2011, Salima, Malawi.

cross-cutting issue, mainstreamed in resource management and development efforts in fishing and fish farming communities. **Box 15** provides a simple checklist for assessing child labour issues when planning support programmes in fisheries and aquaculture.

Poverty and child labour can be addressed by policies and actions promoting integration of decent work concerns as part of development and addressing the issue of interactions between adult and child employment, including youth employment – for those who are above the minimum legal age for employment and have completed compulsory education. This may also include support for safe

and productive migration of youth so they can obtain decent work. Migration awareness campaigns and safe migration services are essential for both sending and receiving communities and locations. Promotion of gender equity is another important part of poverty eradication strategies. Empowerment and creating awareness among communities – men and women, boys and girls – of their rights should be incorporated into programmes addressing poverty and related child labour concerns.

Successful implementation of poverty-focused programmes can turn the vicious cycle of poverty and child labour into a virtuous cycle as shown in **Figure 1**.

Box 16: Junior Farmer Field and Life Schools (JFFLS)

Youth employment is a major concern in many countries around the world. To support decent work and youth employment, FAO introduced the Junior Farmer Field and Life School (JFFLS) approach in 2004. The approach is now used in a number of countries in Africa, Asia (Nepal) and the Near East (Occupied Palestinian Territory[28]) and has benefited both boys and girls.

The JFFLS approach combines agricultural and life skills. Specifically trained extension workers, teachers and social animators use a participatory methodology to pass on agricultural knowledge and life skills to adolescent boys and girls. For one entire school year, a multidisciplinary team of facilitators leads participatory sessions with groups of youth. When working with children and youth attending school, these sessions are given two to three times a week in the field and classroom after regular school hours. The one-year learning programme follows the crop cycle; links are established between agriculture, nutrition, gender equality, child protection, education, business skills, health, hygiene, sanitation and other life-skills knowledge so that young participants learn to grow healthy crops while making informed decisions for leading healthy lives. Participatory field activities include crop selection and cultivation, land preparation, pest management, cultivation of medicinal plants and income generation. The approach also links youth graduates to existing farmers' associations or cooperatives, or supports them to form their own young farmers' associations or group enterprises. This helps facilitate their transition to gainful employment through access to knowledge, inputs, services, financing and marketing. With regard to child labour, a specific training module on child labour prevention in agriculture has been developed jointly with ILO[29] although child and youth protection have always been implicitly part of the JFFLS concept. An innovative aspect of the JFFLS is the way youth are encouraged to develop as people; a school timetable includes cultural activities, such as singing, dancing and theatre. This allows youth to develop confidence while keeping local cultural traditions alive.

The JFFLS process brings together different ministries (such as education, labour, agriculture and trade) as well as farmers' and other stakeholders' organizations, unions and youth associations. These linkages have proved crucial for strengthening the capacities of the public administration and civil society. They have also been fundamental for the institutionalization of the JFFLS approach and the development of mechanisms for addressing rural youth unemployment.

The JFFLS concept has so far been used in agricultural communities, but a similar approach will be developed for fishing and fish farming communities, adjusting the curricula to the realities of these communities. The approach will take into consideration the common problem of overfishing and the potential need to seek employment opportunities outside fisheries, and it will be developed in collaboration with stakeholders.

Source: FAO, 2011b (see also www.fao-ilo.org/fao-ilo-youth/fao-ilo-jffls/en/).

The importance of education[30]

A key strategy for poverty alleviation, long-term sustainable development and reduction of child labour is education. Improved access to quality education is likely to have a positive effect on child labour in the shorter term. Governments should provide compulsory, affordable and quality schooling in fishing and fish farming communities. While still providing basic education, the curricula and school hours could be adapted to suit the particular context and rhythm of fishing communities. The aim should be to allow youth to access decent work opportunities, in the fisheries sector or elsewhere. In areas where overfishing is an issue, vocational training can promote diversification out of traditional fisheries-related professions. Formal education could be combined with appropriate apprenticeship and vocational training programmes. In the agriculture sector, Junior Farmer Field and Life Schools (JFFLS) have

28 FAO, 2010a, Dalla Valle, F. – *Promoting employment and entrepreneurship for vulnerable youth in West Bank and Gaza Strip* (see www.fao.org/docrep/012/i1450e/i1450e00. pdf).

29 FAO, 2010a, JFFLS Facilitator Guide Module: Child Labour Prevention in Agriculture (see www.fao.org/docrep/013/ i1897e/i1897e.pdf).

30 This section is largely based on FAO/IFAD/ILO, 2010.

© ILO / M. Crozet

Democratic Republic of Congo, vocational training certificate

force participation and earnings, and improves their ability to organize in the workplace. Moreover, it increases the likelihood that the next generation of children is sent to school. Promotion of girls' education should be combined with efforts to ensure that there are equal employment opportunities for men and women. In some situations, efforts may be needed to make schools more "girl-friendly", ensuring that schools and transport to and from schools are safe, increasing the number of female teachers and separating boys' and girls' latrines. Cultural reluctance to send girls to school can be overcome by girls-only schools. Another strategy for promoting girls' schooling is to free their time for education by improving rural infrastructure (such as water systems and roads) and childcare services, thus making some of their common tasks less time-consuming.

worked well for farming in a number of countries and a similar approach is being developed for the fisheries and aquaculture sector (see **Box 16**).

Even when education is available, specific incentives - such as school feeding programmes or food-for-schooling - may be necessary to encourage children to attend school. The latter implies that other members of the child's family can also benefit from the food rations provided by the school, since they are taken home. Some countries have cash transfer programmes, applicable on condition that school-age children attend school.[31] Where communities include migrants, it is important to cater for their special needs with regard to, for example, enrolment, semester schedules and transfer certificates. In many cases, infrastructures and incentives to teachers to serve in remote fishing communities are required.

Special incentives are needed to get more girls into schools. Girls' education is particularly beneficial as it leads to lower birth rates and infant, child and maternal mortality rates, and helps protect against HIV/AIDS. Education plays a key role in women's empowerment and in improving their access to decent employment. It increases women's labour

Changing attitudes: corporate social responsibility

When consumer awareness of unsustainable or unethical practices increases, the demand on producers and suppliers to use certain production systems or procedures intensifies. Many markets increasingly require sustainably produced fish, certified by ecolabelling and other certification programmes are becoming more important.[32] Similar demands are also made with regard to the use of fair and socially acceptable production and supply practices. As mentioned earlier (**section 4.2**), the recent FAO Technical Guidelines on Aquaculture Certification refer to the adherence to child labour Conventions and the UN has developed a framework for business engagement which includes spreading responsible business standards and practices.[33] Companies worldwide are increasingly concerned with child labour in their supply chains. Child labour is a potential threat to the sustainability of market positioning and access,

32 See, for example, Sainsbury, 2010.
33 See UN Framework for Business Engagement with the United Nations, available at: www.unglobalcompact.org/docs/news_events/9.1_news_archives/2008_09_24/UN_Business_Framework.pdf.

31 Conditional Cash Transfer (CCT) – see, for example, the World Bank, 2009.

© FAO

Indonesia, father putting safety jacket on his son

and it is seen as inconsistent with company values. Reports of child labour in the supply chain may induce customers and top employee candidates to shy away. Some companies take the concept further and see engagement in social responsibility issues as part of building long-term competitiveness (Genier et al., no date).

This development has resulted in companies using social dialogue, international labour standards and collective approaches linking private businesses to other stakeholders, with companies committing to eliminating the root causes of child labour to effectively address the problem. Promoting and supporting corporate social responsibility initiatives can be a powerful strategy for changing attitudes towards child labour. Promoting the establishment of employers' and workers' – as well as producers' and consumers' – associations and supporting their engagement in such initiatives, may reduce the level of tolerance of child labour. For companies, the

possibility of attracting justice-conscious customers by being branded "child labour free" constitutes an important economic incentive.

Investing in technologies and practices to reduce the demand for child labour

In addition to changing attitudes, the demand for child labour can be reduced by introducing technologies and practices that eliminate the need for children's labour. This could include, for example, improved community infrastructure with regard to water supplies, as well as roads, transport and landing site arrangements, to avoid carrying heavy loads. If children are employed in fishing because they are paid less since profitability is an issue (because of overfishing, decreasing fishery resources and hence yields), addressing fisheries management and and should be a priority introducing more sustainable fishing practices.

7.3 Withdrawing children from child labour

The ILO's experience has shown that parents and families who are given a viable choice prefer to keep children out of the workplace. The simple removal of children from the workplace does not have a significant impact unless it is carried out in the context of a national policy that promotes the rights, welfare and sound development of children and encourages their participation in finding solutions to the problem of child labour.

The withdrawal and subsequent rehabilitation of children engaged in child labour includes:

- identifying children engaged in child labour, especially in hazardous work and other worst forms of child labour;

- removing them from the workplace;

- getting them into school and/or skills training;

- ensuring that there is a viable labour market so that those of working age can obtain employment suitable for their age;

- monitoring to ensure that they do not return to the same workplace or move to a new, or even worse, workplace;

- promoting income-strengthening activities or links to social services for families.

Sometimes immediate action is needed to withdraw children from the worst forms of child labour, linking them to social and educational services, and when necessary provide them with rehabilitation. Measures to withdraw children may rely on persuasion through dialogue with parents, children, employers or law enforcement authorities, but sometimes police-led "rescue" operations may be necessary. Community-based, integrated initiatives tailored to the specific needs of each target group, with close community participation, have proven to be the most effective solutions. For withdrawal to be sustainable parents and families must have alternatives (see **Box 17** for an example from Ghana).

7.4 Protecting children from hazardous work

In some cases, children above the minimum legal age for employment and engaged in hazardous work may be kept in employment if protection measures are guaranteed. Protection entails changing working conditions sufficiently to make them safe for children, or ensuring that children only engage in non-hazardous tasks and activities. Children in the 15–17-year age group, who have reached the legal minimum age for employment but with restrictions on the type of work, can thus be

Box 17: Trafficked children on Lake Volta, Ghana

The International Organization for Migration (IOM) has withdrawn close to 700 children from child labour in the fisheries sector on Lake Volta, Ghana. The children released from child labour have been taken to government rehabilitation centres and given counselling, medical assistance, educational assistance and art therapy for three and a half months.

Project activities have included visits to fishing villages by IOM and its local partners for awareness raising. Likewise, in the villages of origin, IOM has worked with chiefs, parents and other community members to inform them about the dangers of child labour and human trafficking. Assistance has been given to parents to identify income-generating opportunities so that they will not be forced to send away their children again. Fishers who have released children have also received training and been provided with microcredit to enable them to carry out alternative livelihood activities or improve their fishing techniques without using child labour.

Source: IOM website.

helped to make the transition from hazardous work to youth employment.

Protection measures may include appropriate technical and safety training for children and youth prior to working on fishing vessels, in fish processing or on fish farms. Training could be through special schools or programmes (vocational training or apprenticeships) or integrated into school programmes. Training of employers or, for example, adult family members (if children work within the informal household or extended family context), is equally important.

Appropriate personal protective equipment (PPE) for children that significantly reduces or eliminates the risk of certain hazards, such as properly fitted personal flotation devices/life jackets when on fishing vessels, should be provided as and when necessary. However, it should be stressed that hazardous work of children is not an option: children should never be allowed to carry out hazardous tasks. Safety equipment, including PPE, may not provide sufficient protection for young people, who in addition may not use it, or not use it. Workers and children need to be aware of how to improve their safety and what their rights are.

Strategies and measures to protect children should be closely linked to general efforts by governments to improve OSH in the workplace for all. For work onboard fishing vessels, safety at sea improvements and, more generally, all actions to eliminate or diminish the risk of injuries, death or illness at work would benefit children.

SUMMARY POINTS 7

- The ILO classifies child labour actions into three categories: prevention, withdrawal and protection.

- Prevention is the most important approach for addressing child labour and achieving long-term sustainable results.

- Successful implementation of poverty-focused, participatory and integrated programmes can turn the vicious cycle of poverty and child labour into a virtuous cycle leading to sustainable development.

- Making adequate and affordable education available is a key component of a poverty-focused child labour elimination programme. Special incentives may be required to ensure that children attend school, for example, through school feeding programmes or separate schools for girls.

- Changing attitudes, engaging organizations of fishers, fish farmers, fish workers and employers, and other sectoral institutions, promoting corporate social responsibility and introducing technologies and practices to reduce the demand for child labour are other preventive strategies that governments and development partners should engage in.

- Sometimes urgent action is needed to rescue and rehabilitate children in the worst forms of child labour including hazardous work. Close community participation and collaboration are important for sustainable results.

- For children above the minimum legal age for employment (that is in the 15–17-year age group), improved protection can make working conditions safe; this would transform hazardous work of children into youth employment. Onboard fishing vessels, the availability and use of life jackets are particularly important.

India, rural cooperative meeting
© ILO / M. Crozet

8. Closing the data and knowledge gap

8.1 Finding entry points, partners and tools

To implement the policies and related actions discussed above, there is a need to identify strategies and approaches that allow for successful results. There may be limited awareness and knowledge of child labour issues; moreover, the subject matter may also be highly sensitive, provoke stigma and cause uneasiness at both community and government level. When addressing child labour at national and local level, it is therefore crucial to find entry points, partners and tools that are suitable and which work in the particular local context.

8.2 Entry points

Approaching child labour issues by addressing overall OSH concerns is likely to create multiple benefits – for both children and adults. Risk assessments are a key tool for classifying work and identifying hazardous child labour. It was suggested (**section 5.1**) that where general OSH risk assessments have been carried out, these can be used as a starting point for assessing hazards and risks specific to children. If there are systems and procedures in place for government authorities to monitor OSH and for employers to carry out risk assessments in the workplace, this could constitute an entry point for addressing child labour issues. However, it is more likely that such systems and

procedures exist in the formal sector, while child labour is more generally found in the informal economy.

Nonetheless, starting a discussion on OSH may also be a useful approach in the informal economy and might be linked to discussions on more cost-effective and environmentally-friendly work practices. For example, smoking fish in traditional ovens not only creates hazardous smoke, but also consumes more firewood than when using more efficient smoking ovens (see **section 2.3**, *General safety and health in fisheries and aquaculture*, and **Box 18**).

By integrating child labour concerns into initiatives that address this aspect of the fisheries value chain and working closely with fish smokers, several positive results can be achieved, simultaneously affecting the adult fish smokers (who are often women), the children working with them and the community as a whole.

Another entry point for tackling child labour in fisheries is addressing safety at sea in a more general sense. Accidents at sea can lead to child labour, since children may be required to fill in for injured and dead family members, especially in developing countries without a welfare system. In this context, safety at sea courses and other related activities could act as a launching pad for discussion of special risks to children and child labour (see **Box 19**).

If the elimination of child labour were to become a cross-cutting consideration in all development strategies, programmes and actions – in much the same way as gender mainstreaming – it could be easier to address the issue also in situations where there is poor awareness or even resistance to discussing the matter. By approaching it from the "side line" and with a step-by-step approach, more sustainable results may be achieved than if addressed directly and in a way that may be seen as intimidating. There are of course instances of child labour, especially its worst forms, where immediate action is required to withdraw children. However, in order to eliminate all child labour in the future, the issue must gain general recognition, and solutions that may require a longer term engagement need to be considered alongside direct action.

8.3 Partners: engaging organizations of fishers, fish farmers, fish workers and employers, and other sectoral institutions

In addition to the suggestion that child labour concerns become a cross-cutting issue in development, the need to address child labour through participatory and integrated approaches has been stressed several times in this document. When governments, organizations of fishers, fish farmers, fish workers and employers, and other sectoral institutions and development partners intend to widen the scope of child labour interventions (for example, by explicitly taking poverty and education into consideration) at national and local government level, collaboration between agencies that normally have the main responsibility for child labour and the government agency responsible for fisheries and aquaculture is important. As many fishing and fish farming communities have diversified livelihoods, often including agriculture, it is also necessary to work closely with the ministry of agriculture. Other line agencies needed include those responsible for education, health, welfare, and civil and legal protection. Appropriate leadership and coordination mechanisms – promoting awareness raising, collaboration and policy coherence – are needed to ensure that all agencies work together (see **Boxes 20** and **21**).

Outside government agencies and their development partners, important collaborators include organizations of fishers, fish farmers, fish workers and employers, and other sectoral institutions – as well as private sector operators. Child labour needs to be addressed in close collaboration with employers' and workers' organizations. These organizations often have, or could develop, policies and good practice standards, including for child labour. Employers' and workers' organization

---

Box 18: The Ghanaian *chorkor* oven

In Ghana, smoking is the most widely-used method of preserving, processing and storing fish and is the most common activity for women in fishing communities. However, traditional ovens proved inefficient in capacity and fuel usage, resulting in poor quality smoked fish and significant post-harvest losses. More fuel wood than necessary was used, contributing to forest depletion. Women suffered health risks from smoke inhalation, burns and exposure to raw heat. An improved fish smoking oven, developed by FAO and Ghana's Food Research Institute of the Council of Scientific and Industrial Research, was introduced in Ghana in 1969 and it quickly became popular; it is easy and safe to use, has a high processing capacity, uses little fuel wood, results in shorter smoking time and produces high quality smoked fish. The *chorkor* oven has since been introduced and used in many other countries, including Cameroon, Ethiopia, the Gambia, Guinea, Kenya, Lesotho, Nigeria, Sierra Leone, the United Republic of Tanzania, Uganda and Zambia. It can be adapted for use wherever fish-smoking is part of the post-harvest fisheries tradition.

Source: FAO, 2011c.

Box 19: Safety at sea

Safety at sea should be addressed in a holistic and participatory manner. Measures to improve safety are effective only when the motivation to apply them exists. Safety is related to fisheries management; deficient safety may stem from financial constraints caused by, for example, diminishing catches. When profitability is decreasing because of overfishing, little attention may be paid to investments in required equipment or the use of safe practices. There may also be lack of knowledge or limited availability of suitable equipment, training, support facilities and regulatory frameworks. In countries where appropriate regulations, enforcement and training are in place, there has been a measurable (though not always very large) reduction in the annual number of fatalities over the last 15 years.

Source: FAO Safety for fishermen website.

membership tends to come from the formal sector, but there are examples of trade unions welcoming self-employed producers (farmers, fishers) and others as members, for example the General Agricultural Workers Union (GAWU) in Ghana, the Malawian Congress of Trade Unions and the Self-Employed Women's Association (SEWA) in India (FAO, 2011d). As an initial step in a process of intensifying the action against child labour, it may be useful to investigate which government agencies and partner organizations are available and from there build long-term public-private partnerships.

Although in many countries significant numbers of children are engaged in fisheries and aquaculture, existing national policies and structures to address child labour, such as National Action Plans (NAPs) and Child Labour Units of the Ministry of Labour, rarely integrate a targeted sectoral approach. This is aggravated by the fact that small-scale informal fisheries are sometimes beyond the coverage of labour legislation. Stakeholders in the fisheries and aquaculture sector rarely participate in community, district and national action and coordination on child labour. Sectoral policies for fisheries development and management are in most cases child-labour blind, in the sense that their potential impact (either positive or negative) on child labour is not considered.

The International Partnership for Cooperation on Child Labour in Agriculture supports enhanced dialogue, coordination and engagement of agricultural (including fisheries and aquaculture) stakeholders, such as ministries of agriculture, departments of fisheries and community-based fisheries management institutions, with ministries of labour. Awareness, knowledge and shared understanding by players in both labour and agriculture, as well as the use of language and arguments close to agricultural organizations' mandates are key.

Experience demonstrates that agricultural, including fisheries and aquaculture, stakeholders are more likely to become actively involved in child labour work if they perceive:

- food production, income generation and poverty reduction as means of reducing child labour;
- child labour as a threat to future decent work for youth and adults;
- child labour as a long-term threat to sustainable use and management of the natural resource base.

Sectoral stakeholders can bring innovative solutions to address the root causes of child labour based on specific technical knowledge of production processes and technologies.

Stronger engagement of sectoral institutions can promote the integration of child labour issues in fisheries and aquaculture policies. Their involvement can help target fisheries and other agricultural sectors in national child labour policies (such as National Action Plans) and legislation (such as Hazardous Work Lists).

Box 20: Government agency coordination in Brazil

In Brazil, coordination between different government entities was improved in the context of an initiative to strengthen the labour inspection services. By "intelligence" action – labour inspectors collecting and cross-checking information from different agencies involved in the fisheries sector – and strategic planning with regard to coordination of different actors, cases of child labour were discovered and addressed, and the precarious working conditions prevailing on some vessels were improved. During labour inspection operations in the state of Rio de Janeiro in 2010, carried out as a coordinated operation involving the National Coordination of Port and Waterway Labour Inspection (CONITPA) in collaboration with the Navy, the Federal Police, the Ministry of Labour and Employment, the Ministry of Fisheries and the Ministry of the Environment, children were found working as divers untangling nets from motor propellers. These children were referred to the care of the social services and their employers fined.

Source: ILO, 2010b.

Box 21: Integrating child labour in fisheries and aquaculture policy in Cambodia

The Government of Cambodia is an example of success in mainstreaming child labour issues in fisheries and aquaculture policy. The Fisheries Administration, which is part of the Ministry of Agriculture, Forestry and Fisheries, with support from the ILO and FAO, and in collaboration with the Ministry of Labour and Vocational Training, fisheries producers' organizations, workers' organizations and other key national institutions, started a process to enhance awareness and capacity on child labour and its worst forms in October 2011, culminating a few months later in the National Consultation to Combat Child Labour in the Fisheries Sector, which focused on the identification of appropriate strategies and areas of action. The Consultation elaborated a draft National Plan of Action (NPA) on Eliminating Child Labour in the Fisheries Sector of Cambodia. The NPA was officially endorsed by the Ministry of Agriculture, Forestry and Fisheries, and is in line with Cambodia's Strategic Planning Framework for Fisheries 2010–19 and with the National Action Plan on child labour. It outlines the specific steps and overall strategy to be followed by national stakeholders to address child labour in the fisheries and aquaculture sector, indicating specific responsibilities.

The Government of Cambodia has also included child labour elimination targets in fishing communities as part of the 10-year Strategic Planning Framework for Fisheries and incorporated child labour concerns in the Cambodia Code of Conduct for Responsible Fisheries (CAMCODE).

8.4 Tools

There are a number of tools to address child labour – for ensuring adequate policy, legal and institutional frameworks; for deciding what constitutes child labour; for closing the data and knowledge gap; and for taking action. Some have already been discussed (risk assessment in **section 5.1** and the checklist in **Box 15**). Other examples include:

- A mapping exercise including institutional analysis (outlining existing institutional structures, their mandates, their current and planned programmes and their strengths and capacities) provides a basis for seeking collaboration and introducing child labour considerations as a cross-cutting theme in policies, strategies and programmes. Depending on the scope of the particular initiative, the mapping may be required at national level as well as in specific locations.

- In addition to the institutional analysis, it may be necessary to carry out a review of existing policies and legal provisions (for example: What policies exist to support the elimination of child labour? Are existing policies coherent? Have the ILO Conventions been ratified and translated into national policy? Do labour laws adequately cover fisheries and aquaculture

© Nicole Franz

Malawi, children mending nets

operations, including those family-based and informal?).

- Based on policy, legal and institutional reviews, a National Action Plan for addressing child labour can be drawn up, in line with the Decent Work Country Programme where it exists. This should be done through a consultative process, including multistakeholder participation (through workshops) of relevant government agencies, employers' and workers' associations,

and socioprofessional organizations. Specific roles and responsibilities should be given to each partner and the coordination mechanisms in place (or which need to be put in place) indicated, so as to ensure smooth collaboration. Where migration is a significant phenomenon, regional organizations may need to be involved to ensure that national plans adequately reflect regional mobility.

Box 22: Theatre for development (TFD)

Theatre for Development can be used to pass messages, and for education, participatory analysis and other processes when participation is desired but the issues are complicated and delicate or the social setting does not allow to talk about them openly. UNESCO has played a key role in promoting and developing the concept. A first workshop on TFD was held in Zimbabwe in 1970. Today, many development agencies use the technique. More recently, TFD has proved particularly useful for communicating, educating and informing on subjects such as combating HIV/AIDS or promoting gender equality. It could also be an effective tool for creating awareness about child labour.

Source: UNESCO, 2007, referred to in Westlund, Holvoet and Kebe, 2008.

© FAO

Philippines, girls cleaning small sea shells

- At community level, it is possible to raise awareness and promote actions against child labour through participatory assessments, analyses and monitoring, and by adopting different communication methods (radio and TV programmes, public/village meetings and theatre for development [TFD], see **Box 22** and **Box 23**).

- Socio-Economic and Gender Analysis (SEAGA) is an approach elaborated by FAO in partnership with the ILO, the World Bank and the United Nations Development Programme (UNDP) to enhance the capacity of development specialists and humanitarian officers to incorporate socio-economic and gender analysis into development initiatives

Box 23: Child labour monitoring

"One element of child labour programmes that is expressly designed for sustainability is the 'child labour monitoring' (CLM) system. This mechanism was developed as a temporary support to the labour inspectorate in reaching the informal economy workplaces where child labour occurs most frequently and where almost all jobs pose some sort of physical or psychological risk to children. In its simplest form, a three-person team of community members (such as a school teacher, mothers' club member or retired policeman) are given training in how to monitor child labour. They then periodically visit places where children are likely to be working. If they find a child, they report the case to a specially constituted community committee, as well as to the labour inspector or local government authority for follow-up. Depending on the child's situation, the committee will recommend a course of action, e.g. in the case of younger children this is usually removal from the workplace and their placement in an appropriate educational programme; in the case of older youth it may be improvement and monitoring of the work environment; assistance to the family is another option."

Source: IPEC, 2011, p. 56.

and rehabilitation interventions. It is also potentially useful for addressing child labour, in particular with regard to the different situations of boys and girls.

- Just as gender is a cross-cutting issue to be considered in all sector policies, strategies and plans, so should child labour be. Fisheries and aquaculture policy and decision-makers should call on the relevant expertise and request collaboration when drawing up plans for the sector to ensure that child labour is adequately covered.

- To raise awareness of child labour at national (ministerial) level, the government agency in charge of child labour issues may wish to ensure that regular reports on the child labour situation are published. The agency may also organize seminars and field visits for other line ministries focusing on different aspects of child labour.

SUMMARY POINTS 8

- Approaches and strategies must be applied to address child labour, especially in situations of low awareness and where there may be uneasiness regarding the issue.

- Entry points, partners and tools that are suitable and which work in the specific local context are needed.

- Entry points could be, for example, overall OSH assessments and improvement actions and – in fishing – safety at sea.

- Partners are needed at both national and local level. They include different line ministries and government agencies (needed for an integrated approach), socioprofessional organizations and employers' and workers' organizations.

- There are many potential tools in addition to risk assessments and checklists for reviewing child labour in a particular situation. Examples include policy, legal and institutional analyses; drawing up national action plans through participatory workshops; improving knowledge by making reports and events on child labour available; and – at community level – adopting working methods and communication tools (for example, participatory assessments, radio and TV programmes, public and village meetings, and Theatre for Development).

Summary of recommendations

Based on the above discussion, recommendations for addressing child labour in fisheries and aquaculture are outlined below:

Governments should:

- Ratify international Conventions relevant to the elimination of child labour in fisheries and aquaculture.

- Translate international commitments into national legislation. Ensure that national legislation provides full protection for children according to the CRC and supplemented by the ILO Conventions as required (including in the informal economy and with regard to household chores).

- Ensure implementation of child labour legislation through the use of incentives (negative and positive) and enforcement mechanisms.

- Ensure buy-in from communities and those concerned by involving them directly in the planning and implementation of actions against child labour – consult with relevant stakeholders, socioprofessional organizations, and employers' and workers' organizations when formulating policy and defining programmes relevant to child labour (including for the actions listed here).

- Engage sectoral ministries and agencies – ministries of agriculture and fisheries, departments of fisheries and others – to create awareness of child labour and mainstream child labour in sectoral policies, programmes and regulations.

- Use and promote risk assessments, and define Hazardous Work Lists in line with the ILO Conventions Nos 138 and 182, applying criteria adapted to the characteristics of fisheries and aquaculture.

- Review data requirements concerning child labour in fisheries and aquaculture and integrate these needs into existing information collection systems. Improve awareness of child labour at all levels and promote policy coherence.

- Work actively to prevent child labour by addressing poverty and promoting integrated approaches for development and resource and environmental management in fisheries and aquaculture. Mainstream child labour considerations into these processes (make child labour a cross-cutting issue).

- Provide suitable schooling, free of charge, for fishing and fish farming communities. Review curricula and school hours and adjust them to suit the particular needs of the boys and girls of coastal and inland communities. Introduce school feeding programmes or other incentives to attract children to school.

- Ensure coordination among different line agencies as well as with other partners, at both national and local level.

- Collaborate with organizations of fishers, fish farmers, fish workers and employers, and other sectoral institutions to change attitudes towards child labour.

- Support withdrawal of children from trafficking and other worst forms of child labour.

- Promote safety at sea and other protection programmes in fisheries and aquaculture for the benefit of both children and adults.

- Guarantee freedom of association, social dialogue and collective bargaining in fisheries and aquaculture.

Organizations of fishers, fish farmers, fish workers and employers, and other sectoral institutions and the private sector should:

- Strengthen their actions and organization to promote decent work.

- Actively collaborate with governments and their development partners to find practical solutions to prevent and eliminate child labour.

- Work with governments and other partners to conduct risk assessments and identify hazardous child labour (and draw up and periodically revise Hazardous Work Lists).

- Extend membership to self-employed fishers, fish farmers and fish workers as a means to include those working in the informal economy.

- Promote good practice standards including child labour clauses and engage in awareness raising campaigns.

- Adopt policies and codes to eliminate child labour in the sector.

Development partners should assist governments in implementing the above-defined actions and, in particular, should:

- Mainstream child labour in all (fisheries and aquaculture) development projects and programmes.

- Promote training and awareness raising activities on general child labour issues and international policy to stakeholders, including governments and communities.

- Provide support for carrying out risk assessments and for establishing national Hazardous Work Lists (in accordance with the ILO Child Labour Conventions).

- Support development of educational facilities and schools in fishing and fish farming communities as well as school feeding programmes and other incentive mechanisms.

NGOs and development partners at local level should:

- Support awareness raising and the changing of attitudes, as required, with regard to child labour.

- Assist in organizational capacity-building at local level.

- Assess and monitor child labour issues at community level through participatory approaches.

- Support communities, fisher and fish farmer associations, and other local institutions, including those representing children, to know their rights and to have a voice in decision-making.

- Monitor implementation of child labour-related legislation and hold governments accountable for their actions/non-actions.

- Address the root causes of child labour by supporting access to relevant quality education and training, and to safer and adequate technology.

References

Afenyadu, D. 2010. *Child labour in fisheries and aquaculture – a Ghanaian perspective. Background paper prepared for the Workshop on Child Labour in Fisheries and Aquaculture in cooperation with the ILO*, Rome, 14–16 April 2010, p. 15.

Allison, E.H. ; Béné, C. ; Andrew, N.L. 2011. "Poverty reduction as a means to enhance resilience in small-scale fisheries", in R.S. Pomeroy; N.L. Andrew (eds): *Small-scale fisheries management – frameworks and approaches for the developing world*. Oxfordshire, UK, CABI, pp. 216–238.

Allison, E.H.; Seeley, J. 2004. "HIV and AIDS among fisherfolk: a threat to 'responsible fisheries'?", in *Fish and Fisheries*, Vol. 5, pp. 215–234.

Andrees, B. 2008. *Forced labour and human trafficking: a handbook for labour inspectors* (Geneva, ILO), p. 68.

Béné, C.; Merten, S. 2008. "Women and fish-for-sex: Transactional sex, HIV/AIDS and gender in African fisheries", *World Development*, Vol. 36, No. 5, pp. 875–899.

Béné, C.; Macfadyen, G.; Allison, E.H. 2007. *Increasing the contribution of small-scale fisheries to poverty alleviation and food security*, FAO Fisheries Technical Paper No. 481 (Rome, FAO), p. 125.

Ben-Yami, M. 2000. *Risks and dangers in small-scale fisheries: An overview*. Aug. 2000, SAP 3.6/WP.147 (Geneva, ILO). Available at: www.ilo.org/public/english/dialogue/sector/papers/fishrisk/.

Brigham, C.R.; Landrigan, P.J. 1985. "Safety and health in boatbuilding and repair", in *American Journal of Industrial Medicine*, Vol. 8, No. 3, pp. 169–182.

Centers for Disease Control and Prevention. 2010. "Commercial fishing deaths — United States, 2000–2009", in *MMWR*, Vol. 59, No. 27, pp. 842–845.

Dalhousie University. 2012. Intersectoral Working Group Report July 2012, Dalhousie Marine Piracy Project. Available at: marineaffairsprogram.dal.ca/MAP_Projects/PIRACY_Project/.

De Young, C.; Charles, A.; Hjort, A. 2008. *Human dimensions of the ecosystem approach to fisheries: an overview of context, concepts, tools and methods*, FAO Fisheries Technical Paper No. 489 (Rome, FAO), p. 152.

Dey de Pryck, J. 2013. *Good practice policies to eliminate gender inequalities in fish value chains* (Rome, FAO Gender, Equity and Rural Employment Division).

Environmental Justice Foundation (EJF). 2003. *Smash & grab: Conflict, corruption and human rights abuses in the shrimp farming industry* (London, EJF).

Erondu, E.S.; Anyanwu, P.E. 2005. "Potential hazards and risks associated with the aquaculture industry", in *African Journal of Biotechnology*, Vol. 4, No. 13, pp. 1622–1627.

Food and Agriculture Organization of the United Nations (FAO). 2012. *The state of world fisheries and aquaculture (SOFIA) 2012* (Rome, FAO), 209 pp.

–2011a. "Good practices in the governance of small-scale fisheries: Sharing of experiences and lessons learned in responsible fisheries for social and economic development", in COFI, P. 12.

–2011b. "Junior Farmer Field and Life Schools. Gender and equity in rural societies", in Best practices web page. Available at: www.fao.org/bestpractices/content/11/11_04_en.htm.

–2011c. "Post-harvest processing: the Chorkor oven. Fisheries and aquaculture management and conservation", in Best practices web page. Available at: www.fao.org/bestpractices/content/06/06_02_en.htm.

– 2011d. The state of food and agriculture 2010–11 (Rome, FAO), 160 pp.

–2010a. Report of the Workshop on Child Labour in Fisheries and Aquaculture in cooperation with ILO, Rome, 14–16 April 2010, FAO Fisheries and Aquaculture Report No. 944 (Rome, FAO), p. 24.

–2010b. Report of the Inception Workshop of the FAO Extrabudgetary Programme on Fisheries and Aquaculture for Poverty Alleviation and Food Security, Rome, 27–30 Oct. 2009, FAO Fisheries and Aquaculture Report. No. 930 (Rome, FAO), p. 68.

–2007. The state of world fisheries and aquaculture (SOFIA) 2006 (Rome, FAO), 180 pp.

–2005. "World inventory of fisheries. Prevention of emergencies", Issues Fact Sheets. Text by UweBarg, in FAO Fisheries and Aquaculture Department (Rome, FAO). Available at: www.fao.org/fishery/topic/16617/en.

–2003. "Safety at sea. Fisheries and aquaculture topics". Topics Fact Sheets. Text by Peter Manning, in FAO Fisheries and Aquaculture Department (Rome, FAO). Available at: www.fao.org/fishery/topic/12272/en.

–1995. Code of Conduct for Responsible Fisheries. 41 pp.

FAO/IFAD/ILO. 2010. Breaking the rural poverty cycle: Getting girls and boys out of work and into school, Gender and Rural Employment Policy Brief No. 7. 4 pp. Available at: www.fao.org/docrep/013/i2008e/i2008e07.pdf.

Genier, C.; Stamp, M.; Pfitzer, M. No date. Corporate social responsibility in the agrifood sector: Harnessing innovation for sustainable development, prepared for FAO, sponsored by Nestlé, 37 pp. Available at: www.fsg.org/Portals/0/Uploads/Documents/PDF/CSR_in_the_Agrifood_Sector.pdf?cpgn=WP%20DL%20-%20CSR%20in%20the%20Agrifood%20Sector.

IMO/FAO/UNESCO-IOC/WMO/WHO/IAEA/UN/UNEP Joint Group of Experts on the Scientific Aspects of Marine Environmental Protection (GESAMP). 1997. Towards safe and effective use of chemicals in coastal aquaculture, Rep. Stud. GESAMP, No. 65, 40 pp.

Hai, A.; Fatima, A.; Sadaqat, M. 2010. "Socio-economic conditions of child labor – A case study for the fishing sector on Balochistan coast", In International Journal of Social Economics, Vol. 37, No. 4, p 316–338.

Hausen, B.M. 1986. "Contact allergy to wood", in Clinics in Dermatology, Vol. 4, No. 2, p. 65–76.

International Collective in Support of Fishworkers (ICSF). 2011. Indonesian fishermen reduced to scavenging plastic, SAMUDRA News Alert 29 April 2011. Text by Elisabeth Oktofan based on Jakarta Globe article.

–2010. Web page of ICSF International Workshop on Recasting the Net: Defining a gender agenda for sustaining life and livelihood in fishing communities, Chennai, India, 7–10 July 2010. Available at: icsf.net/icsf2006/jspFiles/wif/wifWorkshop/english/about.jsp.

International Labour Office (ILO). 2010a. *Accelerating action against child labour*, Global Report under the follow-up to the ILO Declaration on Fundamental Principles and Rights at Work. International Labour Conference 99th Session 2010 (Geneva), 98 pp.

–2010b. *The good practices of labour inspection in Brazil: the maritime sector*. (Brasilia), 76 pp.

–2010c. Guidelines for port State control officers carrying out inspections under the Work in Fishing Convention, 2007, No. 188, TMEPSCG/2010/12 Sectoral Activities Programme (Geneva), 90 pp.

–2007. Decent working conditions, safety and social protection. Work in Fishing Convention No. 188, Recommendation No. 199, Sectoral Activities Branch, (Geneva), 24 p.

–2007b. *Child trafficking. The ILO's response through IPEC*, 8p.

– 2006. *Tackling hazardous child labour in agriculture: Guidance on policy and practice - User guide*, (Geneva), 323p.

–2002. *A future without child labour*, Global Report under the follow-up to the ILO Declaration on Fundamental Principles and Rights at Work, Report I(B) International Labour Conference 90th Session 2002 (Geneva), 153p.

–2000. *Safety and health in the fishing industry*, Report for discussion at the Tripartite Meeting on Safety and Health in the Fishing Industry, Geneva, 13–17 Dec. 1999 (Geneva). Available at: www.ilo.org/public/english/dialogue/sector/techmeet/tmfi99/tmfir.htm.

–1998. *Report VI (1): Child Labour: Targeting the Intolerable* (Geneva), 123p.

–1999. Tripartite Meeting on Safety and Health in the Fishing Industry, Geneva, 13–17 Dec. 1999.

ILO/IPEC-SIMPOC. 2007. *Explaining the demand and supply of child labour: A review of the underlying theories*, (Geneva, ILO), 53p.

International Partnership for Cooperation on Child Labour in Agriculture (IPCLA). 2011. *Capacity development on child labour in agriculture*, Draft Report. Available at: www.fao-ilo.org/fileadmin/user_upload/fao_ilo/pdf/Report_FAO_ILO_workshop_Malawi_May_2011_Final.pdf.

International Programme on the Elimination of Child Labour (IPEC). 2011. *Children in hazardous work: What we know, what we need to do* (Geneva, ILO), 106p.

IPEC-ILO Cambodia. 2003. *Checklist for self monitoring assessment on occupational safety and health – Fishing sector*, written by Mar Sophea (Phnom Penh, Cambodia), 22p.

Lopata, A.L.; Baatjies, R.; Thrower, S.J.; Jeebhay, M.F. 2005. "Occupational allergies in the seafood industry – a comparative study of Australian and South African workplaces", in *International Maritime Health*, Vol. 55, Nos 1–4, pp. 61–73.

Lugano, A.; Zacharias, C. 2009. "The lake that gives, the lake that takes. Access to health care for fisherfolk at Lake Chilwa, Malawi", Thesis for Master's Degree in International Health, Karolinska Institutet, Stockholm, Sweden, 45 p.

Markkanen, P. 2005. "Dangers, delights and destiny on the sea: Fishers along the east coast of North Sumatra, Indonesia", In *New Solutions*, Vol. 15, No. 2, 126 p.

Mathew, S. 2010. Children's work and labour in fisheries: A note on principles and criteria for employing children and policies and action for progressively eliminating the worst forms of child labour in fisheries and aquaculture. Background paper prepared for the Workshop on Child Labour in Fisheries and Aquaculture in cooperation with ILO, Rome, 14–16 April 2010, 13p.

MEMMEAS/CAB. 2012. Arrêté No. 009, Côte d'Ivoire. Available at: www.ilo.org/dyn/natlex/docs/MONOGRAPH/89333/102599/F2046941423/CIV-89333.pdf.

MLVT/Winrock. 2011. Research report of hazardous child labor in subsistence freshwater fishing sector. General Directorate of Labour, Department of Child Labour. Children's empowerment through education services (CHES) project – Eliminating worst forms of child labour in agriculture. Winrock International, research collaboration. Department of Child Labour. Crossroads to development. May 2011. 72p.

Moreau, D.T.R.; Neis, B. 2009. "Occupational safety and health hazards in Atlantic Canadian aquaculture: Laying the groundwork for prevention", in *Marine Policy*, Vol. 33, pp 401–411.

Njock, J-C.; Westlund, L. 2010. "Migration, resource management and global change: Experiences from fishing communities in West and Central Africa", in *Marine Policy*, Vol. 34, No. 4, July 2010, p. 752–760.

O'Riordan, B. 2006. *Growing pains*, Samudra Report No. 44, July 2006, pp8–13.

Sainsbury, K. 2010. *Review of ecolabelling schemes for fish and fishery products from capture fisheries*, FAO Fisheries and Aquaculture Technical Paper No. 533 (Rome, FAO), 93p.

Sossou, M.A.; Yogtiba, J.A. 2009. "Abuse of children in West Africa: Implications for social work education and practice", in *British Journal of Social Work*, Vol. 39, pp1218–1234.

Tabatabai, H.. 2003. *Mainstreaming action against child labour in development and poverty reduction strategies* (Geneva, IPEC-ILO).

United Nations Office on Drugs and Crime (UNODC). 2006. *Children and drugs. Perspectives No. 1* (Vienna).

Westlund, L.; Holvoet, K.; Kébé, M. 2008. *Achieving poverty reduction through responsible fisheries: lessons from West and Central Africa*, FAO Fisheries and Aquaculture Technical Paper. No. 513 (Rome, FAO), 154p.

Whitman, S.; Williamson, H.; Sloan M.; Fanning, L. 2012. *Dalhousie Marine Piracy Project: Children and youth in marine piracy – Causes, consequences and the way forward*, Marine Affairs Program Technical Report No. 5. Available at Dalhousie University Libraries: http://libraries.dal.ca.

World Bank. 2012. *The hidden harvests – the global contribution of capture fisheries*, Report No. 66469-GLB (Washington, DC), 92p.

–2009. *Conditional cash transfers – reducing present and future poverty*, by A. Fiszbein & N. Schady with F.H.G. Ferreira, M. Grosh, N. Kelleher, P. Olinto & E. Skoufias. (Washington, DC, The International Bank for Reconstruction and Development/The World Bank). Avaiable at: http://siteresources.worldbank.org/INTCCT/Resources/5757608-1234228266004/PRR-CCT_web_noembargo.pdf.

UNAIDS. 2010. *Global report: Joint United Nations Programme on HIV/AIDS (UNAIDS) report on the global AIDS epidemic 2010*, 364p.

UNESCO. 2007. *Theatre and development* (Bureau of Public Information), 2p. Available at: unesdoc. unesco.org/images/0015/001502/150213e.pdf).

Zdunnek, G.; Dinkelaker, D.; Kalla, B.; Matthias, G.; Szrama, R.; Wenz, K. 2008. *Child labour and children's economic activities in agriculture in Ghana.* Centre for Advanced Training in Rural Development, Humboldt Universitätzu Berlin, Faculty of Agriculture and Horticulture. SLE Publication Series S233, 143p.

Websites and Programmes

- ILO Sectoral Activities website: www.ilo.org/sector (then click on "Shipping; ports; fisheries; inland waterways" in the left column).

- FAO web page on child labour in agriculture: www.fao-ilo.org/fao-ilo-child/.

- International Organization for Migration (IOM) website: www.iom.int/.

- ILO International Programme on the Elimination of Child Labour (IPEC): www.ilo.org/ipec/lang--en/index.htm.

- FAO Safety for fishermen website: www.safety-for-fishermen.org.

- International partnership for cooperation on child labour in agriculture (ILO, FAO, IFAD, IFPRI, IFAP, IUF): www.ilo.org/agriculture-partnership.

- FAO Socio-economic and Gender Analysis (SEAGA): www.fao.org/gender/seaga/en/.

- Understanding Children's Work (UCW), an inter-agency research cooperation project on child labour: www.ucw-project.org/.

Appendix 1:
Developmental differences between child and adult workers

The main developmental differences between child workers and adult workers include:[34]

General

- Tissues and organs mature at different rates, and therefore there is not a specific vulnerable age in general. The age at which the child is most vulnerable depends on the hazard and the degree of risk.
- Per kilogram of body weight, children breathe more air, drink more water, eat more food and use more energy than adults. These higher rates of intake result, for example, in greater exposure to diseases (pathogens) and toxic substances/pollutants.
- Given children's small physical size, being asked to do tasks beyond their physical strength may pose additional risks.

Skin

- A child's skin area is 2.5 times greater than an adult's (per unit body weight), which can result in greater skin absorption of toxics. Skin structure is only fully developed after puberty.
- Children have thinner skin, so toxics are more easily absorbed.

Respiratory

- Children have deeper and more frequent breathing and so can breathe in more substances hazardous to health.
- A resting infant has twice the volume of air passing through the lungs compared to a resting adult (per unit of body weight) over the same time period.

Brain

- Maturation can be hindered by exposure to toxic substances.
- Metals are retained in the brain more readily in childhood and absorption is greater (for example, lead and methyl mercury).

---

34 This appendix is based on an earlier draft of ILO, 2006.

Gastro-intestinal, endocrine and reproductive systems and renal function

- The gastro-intestinal, endocrine and reproductive systems and renal function are immature at birth; they mature during childhood and adolescence, therefore the elimination of hazardous agents is less efficient. Exposure to toxic substances in the workplace can hinder the process of maturation.

- The endocrine system and the hormones it generates and controls play a key role in growth and development. The endocrine system may be especially vulnerable to disruption by chemicals during childhood and adolescence.

Enzyme system

- The enzyme system is immature in childhood, resulting in poorer detoxification of hazardous substances.

Energy requirements

- Since children are growing, they have greater energy consumption, which can result in increased susceptibility to toxins.

Fluid requirements

- Children are more likely to dehydrate as they lose more water per kilogram of body weight through:
    - lungs (greater passage of air)
    - skin (larger surface area)
    - kidneys (inability to concentrate urine).

Sleep requirements

- 5–10-year-olds require about 10–11 hours of sleep per night for proper development.
- 10–18-year-olds require about 9.5 hours of sleep per night for proper development.

Temperature

- Children are more sensitive to heat and cold as the sweat glands and thermoregulatory system are not fully developed.

Physical strain/repetitive movements

- Physical strain, especially combined with repetitive movements, on growing bones and joints can cause stunting, spinal injury and other lifelong deformation and disabilities.

Cognitive and behavioural development

- Another key factor is the ability of child labourers to recognize and assess potential safety and health risks at work and to make decisions about them. For younger children this ability is weak.

Children are vulnerable

Other factors which increase levels of risk for children:

- Lack of work experience – they are unable to make informed judgements

- Desire to perform well – they are willing to do extra without realizing the risks

- Incorrect safety and health behaviour learned from adults

- Lack of safety or health training

- Risk from inadequate, even harsh, supervision

- Powerlessness in terms of organization and rights

- Increased likelihood of risk-taking behaviours.

Reduced life expectancy

This concept is difficult to quantify, but the earlier a person starts work, the more premature the ageing that will follow. Some studies indicate that working as a child increases the risk of poor health as an adult (UCW, 2010).

Disabilities

Not only are children likely to acquire a disability as a result of child labour, children who already have a disability may be at greater risk in general. Disabled children are less likely to be in school (a low percentage of disabled children worldwide attend primary school) and more likely to be from poor families since disability and poverty are linked. Disabled people are often among the poorest of the poor. While the data are limited, a study conducted by the ILO on children in the fishing sector in Uganda found that 8 per cent had disabilities (Walakira and Byamugisha, 2008). The same study found that 20 per cent of the children's parents had a disability in their families. Depending on the nature of their impairment, disabled children may also be more vulnerable to safety and health hazards, resulting in more serious impairments or new forms of disability.

Little is known about what happens to child labourers who become disabled as a result of their work, or about disabled children who become child labourers, and currently there is no alternative to informed guesswork. Based on evidence of the situation of children with disabilities in developing countries – evidence which is, at best, sketchy – it is likely that disabled child labourers will face great difficulties finding decent work as an adult and integrating into their communities and society. Their chances of attending school are also likely to be greatly reduced. They may have poor access to orthopaedic or prosthetic services and other health support. Lack of access to education, and the likelihood that they will not be able to read, write or calculate, mean they will have very slim chances of acquiring marketable skills to allow them to rise out of poverty and earn a decent livelihood in adulthood.

Appendix 2:
The Work in Fishing Convention, 2007 (No. 188), and Recommendation, 2007 (No. 199)

The Work in Fishing Recommendation, 2007 (No. 199) states, in Part 1 on Conditions for work onboard fishing vessels, the following with regard to the protection of young people:

7. Members should establish the requirements for the pre-sea training of persons between the ages of 16 and 18 working onboard fishing vessels, taking into account international instruments concerning training for work onboard fishing vessels, including occupational safety and health issues, such as night work, hazardous tasks, work with dangerous machinery, manual handling and transport of heavy loads, work at height, work for excessive periods of time and other relevant issues identified after an assessment of the risks concerned.

8. The training of persons between the ages of 16 and 18 might be provided through participation in an apprenticeship or approved training programme, which should operate under established rules and be monitored by the competent authority, and should not interfere with a person's general education.

9. Members should take measures to ensure that the safety, lifesaving and survival equipment carried onboard fishing vessels carrying persons under the age of 18 is appropriate for the size of such persons.

10. The working hours of fishers under the age of 18 should not exceed eight hours per day and 40 hours per week, and they should not work overtime except where unavoidable for reasons of safety.

11. Fishers under the age of 18 should be assured sufficient time for all meals and a break of at least one hour for the main meal of the day.

The ILO has developed guidelines for labour inspection officers to monitor possible violations of national laws and regulations for implementing the minimum age requirements of the Work in Fishing Convention, 2007 (No. 188) among fishers onboard foreign fishing vessels visiting their ports.[35] According to the guidelines, inspectors can use the following indicative sources of information for such assessments:

*   a crew list, a passport or other official document confirming fishers' birth dates;

*   a work schedule with respect to fishers under the age of 18 to determine hours and nature of work;

*   information on types of work onboard that have been identified as likely to jeopardize the safety of fishers under the age of 18;

---

35 Guidelines for port State control officers carrying out inspections under the Work in Fishing Convention, 2007 (No. 188).

- recent accident reports and safety committee reports to determine whether fishers under the age of 18 were involved;

- interviews, in private, with fishers.

If it is found that there are instances of non-compliance with respect to under age fishers, inspectors should take action, including possible detention of the vessel (see Article 9 of Convention No. 188, and chapter 5 of the ILO port State control guidelines), until the situation is rectified (ILO, 2010c).

Appendix 3:
UN collaboration on safety in fisheries and aquaculture

The ILO, IMO and FAO have jointly prepared a number of voluntary instruments on the safety of fishing vessels and their crews. The purpose of the Code of Safety for Fishermen and Fishing Vessels, Part A, which applies to all fishing vessels regardless of size and type, is to provide information with a view to promoting the safety and health of crew members onboard fishing vessels. In addition to providing information as is necessary for the safe conduct of fishing operations, Part A highlights the duties and responsibilities of the Competent Authorities and skippers, as the owner's representative, to ensure that a minimum age is set for entry into the fishing industry and to ensure that all crew are above the minimum age set by the Competent Authority and that they are physically and medically fit for work onboard a fishing vessel.

The following three FAO/ILO/IMO instruments provide information on the design, construction and equipment of fishing vessels: Code of Safety for Fishermen and Fishing Vessels, Part B (applicable to decked fishing vessels at least 24 metres in length); Voluntary Guidelines for the Design, Construction and Equipment of Small Fishing Vessels (applicable to decked fishing vessels of between 12 and 24 metres in length); and Safety Recommendations for Decked Fishing Vessels of less than 12 metres in length and Undecked Fishing Vessels. In addition to providing comprehensive information on safety standards, all three instruments also specifically refer to the requirement for life jackets in children's sizes onboard fishing vessels.

More recently, the three Organizations have finalized the development of the FAO/ILO/IMO Implementation Guidelines, aimed at assisting competent authorities in the implementation of the above-mentioned Part B of the Code, the Voluntary Guidelines and the Safety Recommendations. This document is currently undergoing an acceptance procedure by the governing bodies of the Organizations.

In addition to these instruments, the Organizations have jointly developed the Document for Guidance on Training and Certification of Fishing Vessel Personnel. This document addresses the issue of minimum age of entry into fisher's training schemes and specifically refers to the mandatory minimum age requirements of the International Convention on Standards of Training, Certification and Watchkeeping for Fishing Vessel Personnel, 1995 (STCW-F 1995), which is a binding instrument that entered into force on 29 September 2012.